Translucent Building Skins

Exploring the design of innovative building enclosure systems (or skins) in contemporary architecture and their precedents in earlier twentieth-century modern architecture, this book examines the tectonics, the history and the influence of translucency as a defining characteristic in architecture. Highly illustrated throughout with drawings and full colour photographs, the book shows that translucency has been and continues to be a fertile ground for architectural experimentation.

Each chapter presents a comparative analysis of two primary buildings: a recent project, paired with a historical precedent, highlighting how architects in different eras have realized the distinctive effects of translucency. The included buildings span a variety of program types, ranging from a single-family residence to a factory to a synagogue. Whether it is Pierre Chareau's glass-lens curtain wall at the Maison de Verre, Frank Lloyd Wright's wall of stacked glass tubes at the Johnson Wax Research Tower, or Peter Zumthor's use of acid-etched glass in a double-skin envelope at the Kunsthaus Bregenz, the included projects each offer an exemplary case study of innovations in materiality and fabrication techniques.

Today, among many contemporary architects, there is an engagement with new technologies, new material assemblies, and new priorities such as sustainability and energy efficiency. A resurgent interest in translucency as a defining quality in buildings has been an important part of this recent dialogue and this book makes essential reading for any architect looking to incorporate aspects of translucency into their buildings.

Scott Murray is an architect and Associate Professor of Architecture at the University of Illinois, USA. He is the author of *Contemporary Curtain Wall Architecture*.

Translucent Building Skins

Material Innovations in Modern and Contemporary Architecture

Scott Murray

Routledge
Taylor & Francis Group

LONDON AND NEW YORK

First published 2013 by Routledge

2 Park Square, Milton Park, Abingdon, Oxfordshire OX14 4RN
605 Third Avenue, New York, NY 10017

Routledge is an imprint of the Taylor & Francis Group, an informa business

First issued in hardback 2020

British Library Cataloguing in Publication Data
A catalogue record for this book is available from the British Library

Library of Congress Cataloging in Publication Data
Murray, Scott (Scott Charles), 1971–
    Translucent building skins : material innovations in modern and contemporary
    architecture / Scott Murray.
        pages cm
    Includes bibliographical references and index.
    (pb : alk. paper) 1. Curtain walls. 2. Transparency in architecture.
    3. Architecture—Technological innovations. I. Title.
    NA2940.M89 2012
    721'.2—dc23   2012007176

ISBN: 978-0-415-68930-4 (hbk)
ISBN: 978-0-415-68931-1 (pbk)
ISBN: 978-0-203-10153-7 (ebk)

Typeset in Avenir
by Keystroke, Station Road, Codsall, Wolverhampton

CONTENTS

PREFACE

It is worth recalling that the development of architectural glass—going back several hundred years—
was directed up until about the first quarter of [the twentieth] century toward transparency, toward
the perfection of an ever larger plate of glass to a point of perfect clarity. Now that transparency as
a see-through characteristic has been technologically mastered, it's hardly surprising to find that it is
no longer the issue. The mastery of the transparent glass pane makes way for more complex and
interesting uses of this marvelous material.

Joan Ockman[i]

As opposed to transparent glass, which allows the direct transmission of incident light rays, translucent
materials can absorb, disperse, and amplify light, creating a hybrid condition which is neither transparent
nor opaque but may embody qualities of both. These materials resolve the seemingly paradoxical combination in a single element of the properties of abundant light transmission, on one hand, and obscuration
of view, on the other. Through various fabrication techniques—sandblasting, acid-etching, ceramic fritting,
laminating, and casting—glass may be transformed from a putatively invisible, transparent surface to a
translucent material with depth and presence. Building envelopes—or skins—that utilize translucency may
do so for a variety of reasons, both practical and poetic: to maximize natural lighting to the building's
interior by day, to outwardly illuminate a building at night, to provide privacy, to evoke a meditative or
serene atmosphere, to infuse the architecture with a sense of mystery or ambiguity.

This book presents a series of architectural projects which explore these possibilities in different ways,
with the goal of deciphering the materials and methods of construction that contribute to each building's
unique skin. Another goal is to analyze the work of contemporary architects who engage the concept of
translucency within a broader historical and conceptual context. To this end, each chapter presents a
comparative analysis of two primary buildings: a recent project, completed within the last 15 years, paired
with a historical precedent, generally from the early to mid-twentieth century, highlighting how architects
in different eras have realized the distinctive effects of translucency. The included buildings span a range
of program types, from a single-family residence, a school, and a factory, to a rare-book library, an art
museum, and a synagogue. The paired projects in each chapter share a common theme (a particular
material, a building type, etc.) and serve to link contemporary visionaries to earlier modern pioneers. This
organizing structure is not intended to claim a linear progression or direct influence of the older project
upon the newer, but rather to make an argument that translucency has been and continues to be fertile
ground for architectural experimentation.

Whether it is Pierre Chareau's glass-lens curtain wall at the Maison de Verre, Frank Lloyd Wright's wall
of stacked glass tubes at the Johnson Wax Research Tower, or Peter Zumthor's use of acid-etched glass in
a double-skin envelope at the Kunsthaus Bregenz, the included projects each offer an exemplary case study
of innovations in materiality and fabrication techniques. As attempted in the following chapters, a thorough
understanding of these projects requires equal attention to the often complex technical assemblies that
constitute each building envelope as well as the remarkable experiential and spatial effects created by its
interaction with light.

Introduction

TRANSPARENCY / TRANSLUCENCY / OPACITY

"No term is more important to modern architecture than 'transparency.'"

Hal Foster[1]

"Transparency is arguably one of the most allusive and illusive tropes of modern architectural discourse."

Eve Blau[2]

"Modernity has been haunted, as we know very well, by a myth of transparency."

Anthony Vidler[3]

"I was never really interested in just transparency. In fact, I was always suspicious of it. The zone I feel very comfortable in is the distance between the translucent and the opaque."

Steven Holl[4]

Although the dominance of transparency as an architectural ideal is well established in the history of modern architecture, there are a series of exceptional and often overlooked modern buildings in which architects eschewed transparency in favor of the more complicated, alternative condition of translucency. It is this influential, though less examined, thread of modernism that this book explores.

Whereas *transparency* is characterized by visual openness and the direct transmission of light, and *opacity* results from the complete blockage or reflection of light, *translucency* is demonstrated by materials that capture, manipulate, and disperse light. In simplest terms, the adjective *translucent* is defined as allowing light to pass through while diffusing it such that objects on the other side are not clearly discernible; it is derived from the Latin *trans*, meaning through, and *lucere*, to shine. The condition of translucency as it occurs in architecture may be thought of as existing at a point along a spectrum, the polar ends of which are transparency and opacity. Thus, translucency is rarely a fixed state and may be perceived as nearly transparent, nearly opaque, or an apparently equal blurring of the two, with such variations dependent upon the material itself (the degree of roughness of a glass surface, for instance) or the ambient environmental conditions (the

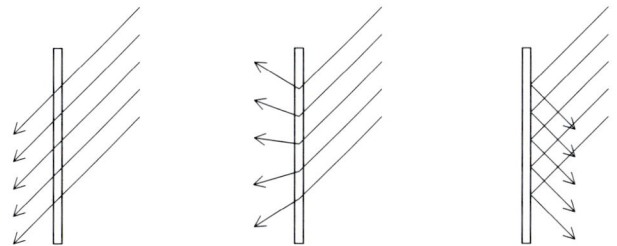

Transparency   Translucency   Opacity

0.1   Interaction of light and material

angle or intensity of sunlight, for example). This idea of translucency, when deployed within windows, walls, and roofs, has, in recent decades, played an important role in expanding concepts of architecture beyond the orthodoxy of modernism (and its dominant association with transparency) and has contributed significantly to current innovations in contemporary architectural design.[5] Today, among many of the contemporary designers whose work is defining early twenty-first-century architecture, there is a renewed interest in the building envelope as a site for experimentation and in translucency as an architectural effect. Although translucency can be deployed within other components—interior partitions, furniture, even floors—it is within the building skin, the interface between interior and exterior space, that it exerts its most significant influence.

As the quotations above suggest, modern architecture has had a fundamental but also complicated relationship with the concept of transparency. The well-documented lineage of this association, as manifested in theoretical texts and built works, can be traced from earlier twentieth-century figures like Sigfried Gideon, Walter Gropius, and Ludwig Mies van der Rohe, to the rise and ubiquity of the 1950s curtain wall, the 1963 critique of literal transparency by Colin Rowe and Robert Slutzky, and the critique of Rowe and Slutzky themselves by subsequent theorists, to the waning influence of transparency (in favor of opacity) during a postmodern interlude, and finally to the resurgence of an interest in glass architecture and optics, as seen, for example, in the work of contemporary architects featured in the 1995 exhibition "Light Construction" at the Museum of Modern Art. Key early works, like Mies's unbuilt Glass Skyscraper Projects (1921–22) and Gropius's Bauhaus Building (1926) in Dessau, Germany, exploited recent developments in frame-structure technology and glass fabrication to envision transparent façades at an unprecedented scale. Writing about his Bauhaus design, Gropius praised transparent glass for provoking "the growing preponderance of voids over solids" and for its "sparkling insubstantiality . . . the way it seems to float between wall and wall, imponderably as the air."[6] In his 1929 book, *Glass in Modern Architecture*, Arthur Korn announced that "the contribution of the present age is that it is now possible to have an independent wall of glass, a skin of glass around a building," which he hailed as "the disappearance of the outside wall."[7] To a greater degree than many of his contemporaries, Mies recognized the complex visual properties of the new glass architecture, noting that "the important thing is the play of reflections and not the effect of light and shadow as in ordinary buildings."[8] Mies's Tugendhat House (1930) in Brno, Czech Republic, displays a unique combination of transparency and translucency in its extensive glazing, and is in that sense a precursor to his later design for Crown Hall (1956) in Chicago. At the Tugendhat House, a long row of transparent floor-to-ceiling windows opens the main living space to a garden. In fact, two of these large windows can be fully lowered into a slot in the floor, thus literally achieving Korn's metaphor of the "disappearance of the outside wall." Mies used equally large panes of translucent glass to enclose the entry hall and stair, providing the desired degree of privacy while creating a carefully choreographed transition from inwardly to outwardly focused spaces.

In their influential 1963 essay, "Transparency: Literal and Phenomenal," Colin Rowe and Robert Slutzky make a distinction between "literal" (or actual) transparency—the ability to see through a material—and the more abstract "phenomenal" (or illusionistic) transparency, in which the layering of planes and volumes suggests the interpenetration of partially revealed, partially hidden spaces beyond.[9] Among other examples, the authors contend that literal transparency is manifested in Gropius's Bauhaus Building while phenomenal transparency is represented by Le

0.2 Bauhaus Building, Dessau, Germany. Walter Gropius, 1926.

Corbusier's Villa at Garches, with a bias clearly expressed toward the latter. The essay proposed a new understanding of transparency as "spatial stratification" and was quickly received as a critique of the accepted notion of transparency and its dominance

0.3 Entry hall. Tugendhat House, Brno, Czech Republic. Ludwig Mies van der Rohe, 1930.

in the mainstream architecture of the time. However, although Rowe and Slutzky present these two types of transparency in binary opposition, they are neither mutually exclusive nor exhaustive. In fact, many of the qualities they attribute to phenomenal transparency would find a closer fit within an expanded definition of translucency, which forgoes literal transparency but still reveals spatial depth and layering in abstraction. Rowe and Slutzky's essay has been skillfully critiqued by Detlef Mertins, who calls it a "reductive and restrictive interpretation" of transparency.[10] Mertins correctly points out the limitations of their exclusively formal analysis of phenomenal transparency, relying as it does on a two-dimensional, frontal reading of a building's façade from a singular fixed viewpoint, as if viewing a painting instead of experiencing a three-dimensional spatial construct.[11]

The related concepts of transparency and translucency are also of course inextricably linked with the phenomenon of light, the luminous energy that has influenced architectural design for

0.4 Literal transparency. Design Research Store, Cambridge, Massachusetts, USA. Benjamin Thompson and Associates, 1969.

centuries.[12] Early examples of translucent building skins can be found in such diverse instances as the soaring stained-glass walls of Gothic cathedrals, like the radiant Sainte Chapelle (1248) in Paris and the Choir at Aachen Cathedral (1414), and the white paper shoji screens of traditional Japanese architecture, like those of the Shoren-in Temple, built in the late thirteenth century in Kyoto (rebuilt in 1895).[13] Many leading architects of the modern movement acknowledged the great theoretical significance of light to our experience of the built environment. Le Corbusier famously defined architecture as "the masterly, correct and magnificent play of masses brought together in light."[14] Louis Kahn declared that "no space, architecturally, is a space unless it has natural light."[15] Steven Holl, a central figure in the contemporary discourse on translucency, writes that "an attention to phenomenal properties of the transmission of light through material can present poetic tools for making spaces of exhilarating perceptions."[16] Holl also observes that "light is for space what sound is for music."[17]

In addition to recognizing the qualitative and experiential impact of light and its interaction with architecture, this discussion also requires an understanding of the quantitative, technical aspects of light. The solar spectrum—what we call sunlight—actually consists of three distinct categories of light distinguished by differences in wavelength. Only one of these types, known as visible light, is perceptible to the human eye. Visible light occupies the middle of the spectrum (with a wavelength range of 380–780 nanometers). The other two types are ultraviolet light, found at the low end of the spectrum (300–380 nm), and infrared light, at the high end (790–3,000 nm). Visible light constitutes about 47 percent of the solar spectrum, while infrared is about 51 percent and ultraviolet light is just 2 percent. Each of these types is relevant to the design of building envelopes, but for different reasons. Infrared light converts to heat when it is absorbed by a material and therefore has implications for thermal performance. Ultraviolet light can have a potentially damaging effect on materials like fabrics and plastics and must therefore be blocked in certain situations. Visible light, however, obviously has the greatest impact when considering the association of light with the architectural conditions of transparency and translucency.

0.5 Shoren-in Temple, Kyoto, Japan. Late thirteenth century, rebuilt 1895.

The physicist Richard Feynman has written about "the truly strange behavior of light" and its interactions with matter. His description of glass as a "terrible monster of complexity" acknowledges the many varied ways that glass can disturb or manipulate photons (particles of light) through partial reflection and refraction.[18] It should be noted that no architectural material of actual substance is truly and fully transparent—a single sheet of clear float glass, for instance, in its basic form (without coatings or other treatments) typically transmits about 92 percent of the visible light striking its surface and reflects the other 8 percent. Such material, however, is normally perceived in application as fully transparent, depending upon environmental conditions and perspective. The light rays that pass through transparent glass do so relatively undisturbed. This contrasts with a translucent material, such as a sheet of sandblasted or acid-etched glass. The surface of this glass, which has been physically roughened through the application of sand or acid, essentially consists of a continuous network of faceted planes that act at the micro-scale as lenses that transmit light particles but divert them from their original paths. Through etching, the surface also acquires a matte finish and therefore reduces the reflective tendencies of the glass. Light becomes complicated; through refraction it becomes diffuse. When used in a building envelope, regular clear glass tends to oscillate in perception between transparency and reflectivity; translucent glass, particularly when rendered with a matte surface, tends to oscillate between translucency (when a light source is present on the other side) and opacity. The oscillation of transparent glass can be experienced at the Glass Pavilion (2006), designed by SANAA for the Toledo Museum of Art in Toledo, Ohio, where, depending upon environmental conditions, the tall double-skin walls of clear glass alternately provide a direct view through the wall or produce a multiplication of reflections that fills the glass surface with distorted imagery. The oscillation of translucent glass is apparent at the Nelson-Atkins Museum of Art (2007) in Kansas City, Missouri, by Steven Holl Architects, where the translucent channel-glass building skin can appear as opaque as stone or may glow brightly from within, depending upon the presence or absence of internal lighting.

When contemplating these issues of light and translucency, the

0.6 Glass Pavilion, Toledo Museum of Art, Ohio, USA. SANAA, 2006.

0.7 Nelson-Atkins Museum of Art, Kansas City, Missouri, USA. Steven Holl Architects, 2007.

0.8 Nelson-Atkins Museum of Art, Kansas City, Missouri, USA. Steven Holl Architects, 2007.

architect is obliged to consider a series of interesting dualities. The first, already mentioned above, is the necessity to engage two modes of architectural performance: the physical/technical properties of materials and their related phenomenal/experiential effects. A second duality relates to the passage of time and the diurnal cycle: translucent building skins often transform dramatically from day to night, creating cyclical variation in the character of interior and exterior spaces. And this leads to yet another: the duality of nature and artifice, relating to the source of light. Translucent façades typically transmit sunlight inwardly by day and electrical lighting outwardly at night.

Translucency is often defined by what it is not. It clearly stands in emphatic contrast with the dominant modernist theme of transparency, as seen in the buildings included in this book. Whereas transparent materials immediately reveal the space beyond, translucent materials do not. However, nor do they completely conceal the space beyond, as opaque materials do. Translucent materials are neither one nor the other, and for this reason, translucency is often identified with a sense of mystery or ambiguity. The architectural theorist Juhani Pallasmaa has written eloquently of the allure of ambiguity:

Mist and twilight awaken the imagination by making visual images unclear and ambiguous: a Chinese painting of a foggy mountain landscape, or the raked sand garden of Ryoan-ji Zen Garden give rise to an unfocused way of looking, evoking a trance-like, meditative state. The absent-minded gaze penetrates the surface of the physical image and focuses in infinity.[19]

In this sense, ambiguity can generate a productive complication of sensory experience. Architects who engage these ideas in their work often do so through the deliberate use of innovative cladding materials and fabrication techniques, resulting in building skins imbued with an enigmatic ambiguity. These constructions respond to the most subtle changes in lighting conditions, and, in contrast to transparency, fully reveal their nature only through the passage of time and through the exploration and experience of both interior and exterior space.

1
Solidified Light

MAISON DE VERRE, PARIS, FRANCE, 1932

HIGGINS HALL INSERTION, NEW YORK CITY, USA, 2005

The related conditions of translucency and luminosity have been definitive themes in the work of Steven Holl, who writes, "The idea of trapping light or building out of blocks of light is something I've long been obsessed with."[1] Indeed, Holl's body of work, dating back to the late 1970s, clearly shows the numerous results of this fascination.[2] More recent designs for large-scale cultural institutions built around the world—beginning with the Kiasma Museum (1998) in Helsinki and including the Nelson-Atkins Museum (2007) in Kansas City and the Nanjing Sifang Art Museum (2012) in China—have focused, in part, on the development of innovative glass building skins that deliver the "trapped light" of Holl's obsession.[3] However, a smaller and perhaps lesser known project, located in Holl's home territory of New York City, offers a particularly instructive case study of this pursuit. The Higgins Hall Insertion (2005) in Brooklyn, designed by Steven Holl Architects, houses the Pratt Institute School of Architecture and includes a distinctively translucent building envelope that infuses the interior with natural light and glows from within at night. It is also the project which most directly recalls an earlier twentieth-century building that has undoubtedly been influential to the work of Holl and other contemporary architects similarly engaged with translucency—Pierre Chareau's Maison de Verre of 1932. This chapter will present an analysis of the unique enclosure systems of these two buildings, their material innovations, and the ways in which the condition of translucency is deployed to address concerns both practical and poetic.

Pierre Chareau had been primarily an interior decorator and furniture designer when he was commissioned in 1928 by Dr. Jean Dalsace and his wife, Annie, to design what would become Maison Dalsace, eventually known as the Maison de Verre, a house clad almost entirely in translucent glass.[4] It would also be the defining project of Chareau's career. Designed in collaboration with the architect Bernard Bijvoet, the Maison de Verre remained relatively obscure in the years following its completion in 1932 but has since come to be recognized as a canonical, yet still somehow estranged, work of modernism. *New York*

1.1 Higgins Hall Insertion, Brooklyn, New York, USA. Steven Holl Architects, 2005.

1.2 Maison de Verre, Paris, France. Pierre Chareau and Bernard Bijvoet, 1932.

1.3 Maison de Verre, Paris, France. Pierre Chareau and Bernard Bijvoet, 1932.

Times critic Nicolai Ouroussoff writes that the house "reflects the magical promise of twentieth-century architecture," and "for architects it represents the road not taken: a lyrical machine whose theatricality is the antithesis of the dry functionalist aesthetic that reigned through much of the twentieth century."[5] The lyrical and theatrical qualities of the house are generated in large part by an innovative building-envelope system and its interactions with light.

In 1928, the Dalsace family acquired a three-story eighteenth-century house in a Left Bank residential neighborhood in Paris. The house shared a masonry party wall with each of two adjacent buildings on its north and south ends and was set back from the street behind a gate, with a paved front-entrance courtyard on the west side and a garden to the east. The new owners intended to demolish the old house and build a new one in its place. However, the top floor apartment was occupied by an elderly tenant who refused to leave and, by law, could not be evicted, so Chareau devised a plan to remove the lower two levels while leaving the top floor intact, supported by a new steel structure beneath, and to insert the new house in the space below. Chareau was actually able to fit three new levels into this space. The program included a medical clinic for Dr. Dalsace's practice on the ground floor, a kitchen, dining room, and double-height living room on the second floor, and bedrooms on the third floor. The design is notable for many reasons: its sensitive choreography of public and private functions, its rich variety of interior spatial conditions, and the elaborately kinetic and modular nature of its furnishings and fixtures (also designed by Chareau). But beyond these, it is Chareau's innovative and transformative use of industrial materials that make the house an important milestone in modern architecture.

The Maison de Verre consists of a structural frame of riveted steel columns and beams, exposed to view on the interior, that carry the concrete floor slabs of the house. The floors cantilever beyond the columns to support the enclosing walls: a continuous grid of translucent glass lenses framed by a network of narrow steel mullions painted black. This system forms both of the building's main façades: one facing to the courtyard and the other to the garden. The house appears like a glass box inserted into

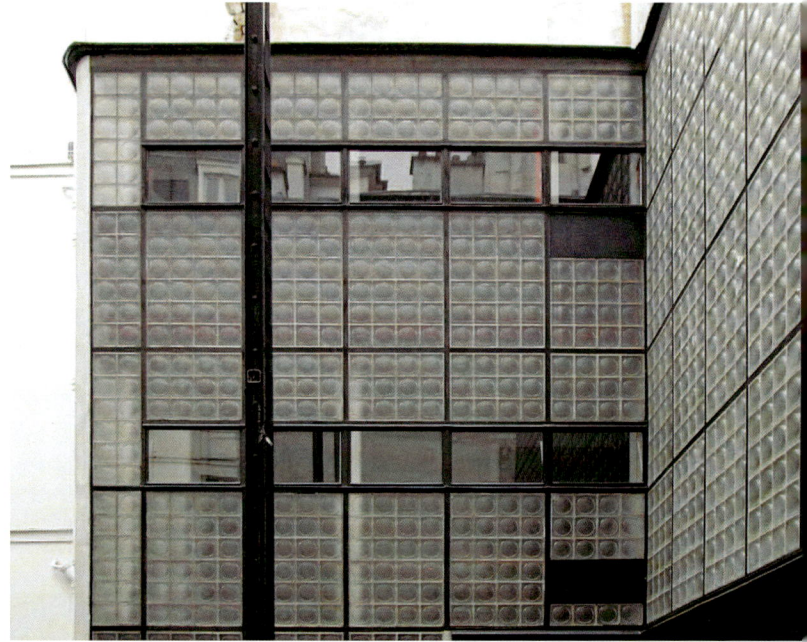

1.4 Maison de Verre, Paris, France. Pierre Chareau and Bernard Bijvoet, 1932.

the existing fabric of traditional masonry buildings that surround it in plan and section. Pierre Chareau described his design process matter of factly, writing:

> I had to build between two party walls, and the plans called for a division of space according to the needs and tastes of modern living habits. There was only one way to get the maximum of light: build entirely translucent facades . . . we began looking for elements which, once assembled, could make unlimited surfaces.[6]

The constraints of the site and his client's program thus led Chareau to develop an enclosure system that would maximize natural light within the house while still providing the privacy required by its medical and residential uses. In contrast to the concentrated sunlight and direct views that would be transmitted through regular transparent glass, the translucent glass walls

1.5 Maison de Verre, Paris, France. Pierre Chareau and Bernard Bijvoet, 1932.

effectively obscure vision while ensuring that sunlight is diffused and distributed more evenly to the interior. An appreciative Dr. Dalsace later wrote that "Chareau has performed the extraordinary feat of building three floors full of light."[7] The effect is of a consistently glowing surface, and the walls essentially become a new kind of continuous light fixture. At night, as internal electrical lights are turned on, the house becomes a lantern as well as a projection screen for hazy shadows that convey through gradation a sense of the interior spatial depth and activity beyond the glass. Chareau apparently also wanted the glass walls to glow inwardly even at night; a series of external floodlights are aimed at the façade, including several mounted to two free-standing steel ladders (installed solely for this purpose) that rise from the courtyard to the full height of the building. This unexpected illumination from an unseen source enhances the sense of mystery provided by the translucent walls. The significance of Chareau's innovative façade design is summed up succinctly by Kenneth Frampton, who writes, "The more or less continuous translucent covering of the Maison de Verre at one stroke does away with the counterpoint between solid and void which one finds in all architecture including the Modern Movement."[8]

The glass units of the Maison de Verre façade have been commonly but erroneously described as glass blocks, which in current terminology implies a thick, hollow, double-walled brick of glass that can be stacked to form a stable wall without additional framing support. Such glass blocks did not come into widespread use until shortly after the Maison de Verre was built (see Chapter 4 for more on glass blocks). In the form used at Maison de Verre, the units are more precisely termed glass *lenses* or *tiles*, which are thin, single-walled, monolithic pieces of cast glass. Their frontal dimensions and general appearance can be similar to glass blocks, but they require different techniques for installation. Frampton has noted that before the Maison de Verre, such glass lenses had only been used once as a primary protective building skin, at Bruno Taut's Glass Pavilion of 1914 in Cologne, Germany.[9] Aside from these two unique examples of continuous glass-lens walls, the material had been more commonly used for utilitarian purposes in discrete, limited areas, such as for storefront clerestory windows, or in horizontal applications, inserted into

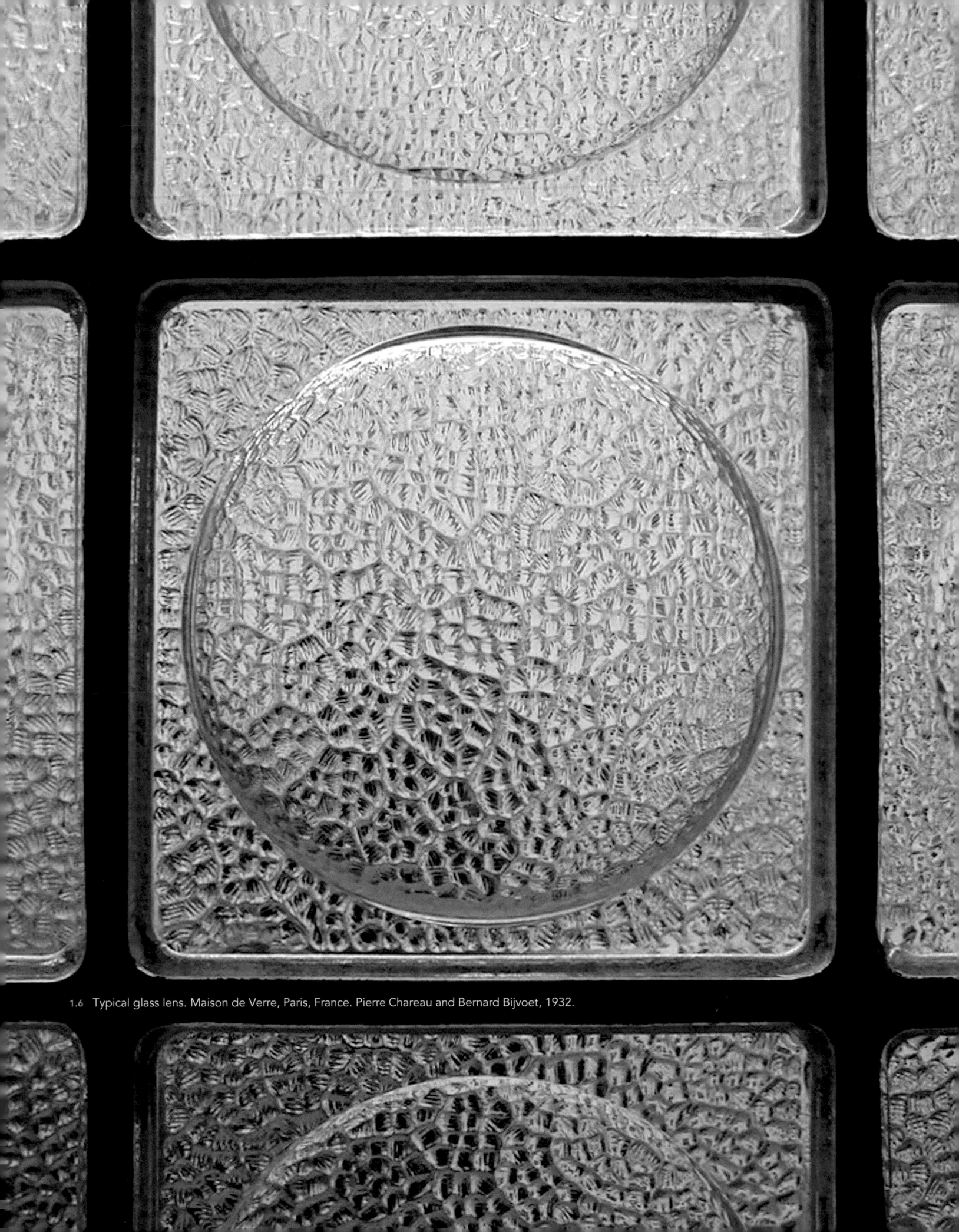

1.6 Typical glass lens. Maison de Verre, Paris, France. Pierre Chareau and Bernard Bijvoet, 1932.

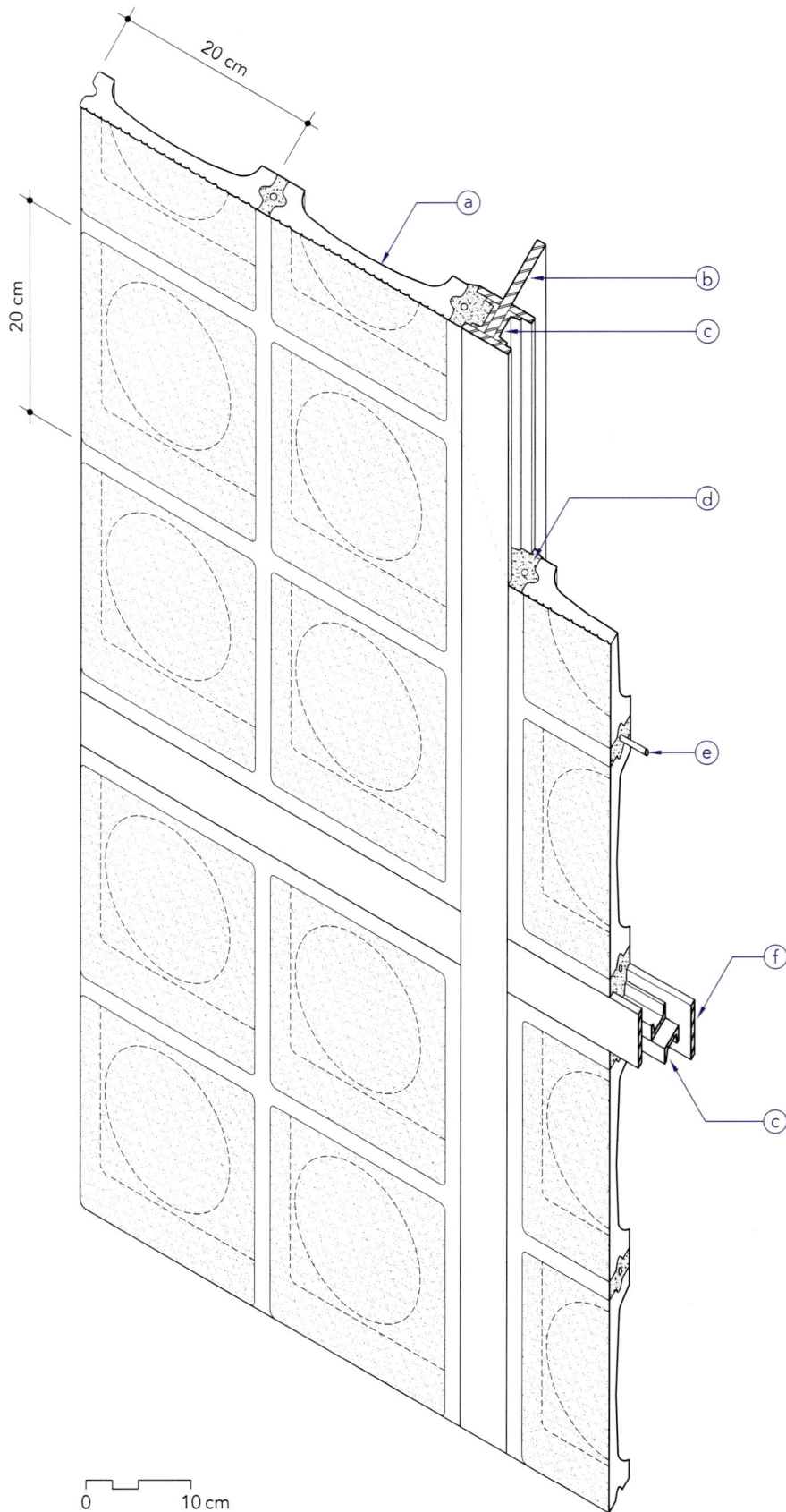

20 cm

20 cm

(a) Translucent glass lens

(b) Steel plate mullion

(c) Steel channel

(d) Mortar

(e) Steel rod reinforcement

(f) Steel plate

0   10 cm

1.7 Glass-lens curtain wall. Maison de Verre.

1.8 Typical module. Maison de Verre, Paris, France. Pierre Chareau and Bernard Bijvoet, 1932.

concrete floor slabs to allow light to pass through to spaces below.

The glass unit that Chareau selected for the Maison de Verre façade was known as the Nevada lens, a product of the French glass-manufacturing firm Saint Gobain. Each lens measures 20 centimeters square (7-¾ inches) with a thickness of 4 centimeters (1-½ inches). The outside face (as positioned at the Maison de Verre) has a random, cellular orange-peel texture physically imprinted on the glass surface. The inside face of each lens contains a single circular, concave indention that is visually discernible from both sides and creates a distinctive pattern on the façade. The overall translucency of the lens derives from the combination of these textures and the resulting refraction of transmitted light.

Given the thin profile of the glass lenses and their inherent instability when oriented vertically, it would have been impractical to stack them to the full height of the façade. To address this issue, Chareau developed a repeating module, four lenses wide by six high, framed vertically and horizontally by steel mullions. The mullions support the gravity loads of the lenses at the base of each module and also transmit lateral wind loads to the floor and roof slabs beyond (but are not otherwise structural themselves). Vertical mullions consist of a flat steel plate measuring 9 by 100 millimeters (⅜ × 4 inches) in cross section, flanked by two steel channels each measuring 15 by 30 millimeters (⁹⁄₁₆ × 1-³⁄₁₆ inches).[10] Horizontal mullions span between the vertical mullions and comprise two steel channels with dimensions similar to those in the verticals. The joint between individual glass lenses is filled with mortar, into the center of which is embedded a steel reinforcing rod, 5 millimeters in diameter (¼ inch). Chareau's innovative and untested design for an entire façade of glass lenses apparently exceeded even the comfort zone of their manufacturer. Saint Gobain produced and supplied the Nevada lenses (more than 3,600 in total) but declined to offer a warranty ensuring that the wall system would remain weatherproof.[11]

Beyond the quality of light provided, Chareau's design establishes a deliberate relationship between the modularity of the glass walls and the interior experience of the house. The dimension of the typical four-lens-wide façade module—about 91 centimeters (3 feet)—is repeated inside as the width of doors,

1.9 Interior. Maison de Verre, Paris, France. Pierre Chareau and Bernard Bijvoet, 1932.

staircases, and even furniture pieces, lending a dimensional unity to the whole. The repeating circular motif of the lenses is echoed by the raised circular pattern of the rubber-tile flooring. In order to facilitate natural ventilation and the occasional view to the exterior, horizontal bands of steel-framed operable windows with transparent plate glass are incorporated into the façade grid, mainly in the bedrooms and the kitchen. In anticipation of the poor insulating performance of the façade's single layer of glass, Chareau included a small semicircular gutter integrally cast into

1.10 Maison de Verre, Paris, France. Pierre Chareau and Bernard Bijvoet, 1932.

1.12 Maison de Verre, Paris, France. Pierre Chareau and Bernard Bijvoet, 1932.

the floor slab at the base of the glass walls to collect and control any condensation that might form on the cold surface of the glass.

In his 1943 book, *The Place of Glass in Buildings*, John Gloag writes, "Glass is a revolutionary material: it enables the architect to fabricate and manipulate something that is universal but impalpable, beyond price but without cost—daylight."[12] The unique characteristics of this material—particularly its apparent ability to *fabricate* light, in Gloag's terms—are in fact intensified when glass is rendered translucent, and its physical manipulation of light rays results in an apparent amplification of light. This phenomenon can be clearly seen in Pierre Chareau's Maison de Verre and is equally evident in Steven Holl's Higgins Hall Insertion, a project that occupies similar conceptual terrain and addresses analogous issues from a twenty-first-century perspective.

1.11 Maison de Verre, Paris, France. Pierre Chareau and Bernard Bijvoet, 1932.

* * *

1.13 Higgins Hall Insertion, Brooklyn, New York, USA. Steven Holl Architects, 2005.

Although it was built seven decades later on another continent, the Higgins Hall Insertion shares several uncanny similarities with the Maison de Verre. While one is a school and the other a residence, both buildings are three-story additions, sheathed in glass and carefully placed between much older existing structures. Each building is set back from the street, facing an entry courtyard on one side and a garden on the other. Each building uses translucency to establish a luminous counterpoint to the opacity of adjacent load-bearing masonry structures. In each, a continuous translucent glass skin is strategically punctured by steel-framed transparent windows for views and ventilation. And each building exploits glass construction materials traditionally used in more utilitarian, industrial applications (channel glass at Higgins Hall, glass lenses at Maison de Verre) to create elegant and functional architectural statements widely recognized as important contributions to the field.[13]

The School of Architecture at Pratt Institute in Brooklyn, New York, was housed in Higgins Hall, a late nineteenth-century building shaped like an H in plan, when a 1996 fire destroyed the center section (connecting the two ends of the H). Steven Holl Architects were hired to design a new addition (or *insertion*, as it would come to be known) to replace the missing piece. Holl's office worked in collaboration with Rogers Marvel Architects, who were also overseeing the renovation of the two remaining sections of the original building. Construction of the insertion was delayed by seven years due to unanticipated structural and water-damage issues in the existing building and was not completed until 2005.[14]

1.14 Higgins Hall Insertion, Brooklyn, New York, USA. Steven Holl Architects, 2005.

1.15   Higgins Hall Insertion, Brooklyn, New York, USA. Steven Holl Architects, 2005.

Holl's design strategy was to introduce an unapologetically contemporary focal point to the existing architectural composition that would provide a new main entry sequence to the complex and once again link the two newly isolated halves. The critic Fred Bernstein writes that "the lines between Pratt's nineteenth-century architecture and Holl's twenty-first-century contribution couldn't be clearer—the result being that neither undermines the other."[15] The insertion contains 2,000 square meters (22,500 square feet) of space, including a ground-floor lobby and exhibition hall that leads down to a below-grade auditorium and up to two levels of studio space. Holl's design draws upon the fact that the floor levels in the two existing structures to which it connects are out of alignment, differing in elevation at the ground floor by only 12 millimeters (½ inch) but at the third floor by 1.45 meters (4 feet, 9 inches). These misaligned levels are extended from each side

to the center of the addition, where an internal promenade ramp joins them. Holl terms this location the "dissonant zone," and marks it on each façade with an asymmetrical patchwork of differently sized windows.[16] In contrast to the heavy load-bearing masonry of the existing building, the addition's structure is a frame of precast concrete columns and beams, enabling the use of an all-glass curtain wall on the front and rear façades. The structural elements are all left exposed on the interior to serve a didactic role for architecture students.

The curtain-wall system consists of two distinct parts: the majority of each façade is covered with vertically oriented translucent channel-glass planks that span continuously from floor to floor, interrupted only by the "dissonant zone," with its irregular aggregation of steel-framed transparent glass panels, some of which hang vertically while others are tilted slightly out of plane.

1.16 Studio interior. Higgins Hall Insertion, Brooklyn, New York, USA. Steven Holl Architects, 2005.

The mullions of the steel windows, as well as the horizontal aluminum supports of the channel-glass system, are painted a deep oxide red, matching the brick color of the existing buildings. The selective inclusion of transparent glass adjacent to the interior ramp provides a moment of pause and reflection, reorienting the occupant to the outside world. The translucent surfaces, on the other hand, focus the building inwardly, giving only a hint of exterior conditions through the shifting intensity of sunlight and the shadows of passing clouds. The interior is infused with natural light during the day, and at night, when the studio spaces are often full of students at work on their projects, interior lighting causes the building to glow, illuminating the public space of the entrance courtyard and displaying through moving shadows the activity inside.

Channel-glass walls, consisting of interlocking U-shaped planks of translucent glass, were a relatively new proposition in the USA at the time of Holl's design. However, channel glass has been used for several decades in Europe, most commonly for industrial

1.18 Interior. Higgins Hall Insertion, Brooklyn, New York, USA. Steven Holl Architects, 2005.

rigid channel profile. The basic ingredients of channel glass are similar to standard float glass—sand, soda, lime, and recycled glass. These materials are melted in a furnace and then drawn out to form a continuous ribbon of molten glass that is then passed through a series of steel rollers to form the channel shape and to impart the desired surface texture. After cooling in an annealing lehr, the ribbons are cut to the required length. The channel-glass pieces may then also be tempered (as they are at Higgins Hall) if needed for additional strength and impact resistance.

Holl's office worked with the building-envelope consultants of R.A. Heintges & Associates to develop the material palette and detailing strategies for the curtain wall, which was installed by local firm W&W Glass Systems.[17] The channel-glass system used at Higgins Hall was manufactured by the German company Glasfabrik Lamberts, whose products are distributed in the USA by Bendheim Wall Systems. The face width of each plank is about 260 millimeters (10-¼ inches) with side flanges of 60 millimeters (2-⅜ inches). The glass is 6 millimeters thick (¼ inch). Each surface is imprinted with a random orange-peel texture, similar to that of the Maison de Verre lenses, which makes the glass translucent and matte. The channel-glass planks run vertically from floor slab to floor slab, where they are supported at their top and bottom edges by horizontal extruded-aluminum channels approximately 65 millimeters high by 100 deep (2-½ by 4 inches) in profile. The planks are arranged in a nested double layer—two rows of planks spaced about 75 millimeters (3 inches) apart with their flanges facing into the cavity created between them. The joints between the planks are typically 6 millimeters (¼ inch) wide and are filled with clear silicone sealant for weatherproofing. Although channel glass can be used in a single layer, the double-glazed configuration, with its buffering air cavity, helps improve the thermal insulating properties of the wall. This performance is further enhanced at Higgins Hall by one of the first US installations of translucent capillary insulation encapsulated within the channel-glass cavity. The insulation, known by the trade name Okapane, is in panels consisting of fine, honeycombed, clear acrylic tubes faced with a glass-fiber tissue. In addition to providing thermal and acoustic insulation, Okapane further diffuses light passing through the assembly.

facilities though more recently in cultural and institutional buildings as well. The main advantages of this material are, first, its light-diffusing properties resulting from a surface texture that renders the glass translucent, and, second, its ability to span much longer distances than regular flat glass, up to 6 meters (20 feet) or more without needing additional mullions, thanks to its structurally

1.19 Channel-glass wall assembly. Higgins Hall Insertion.

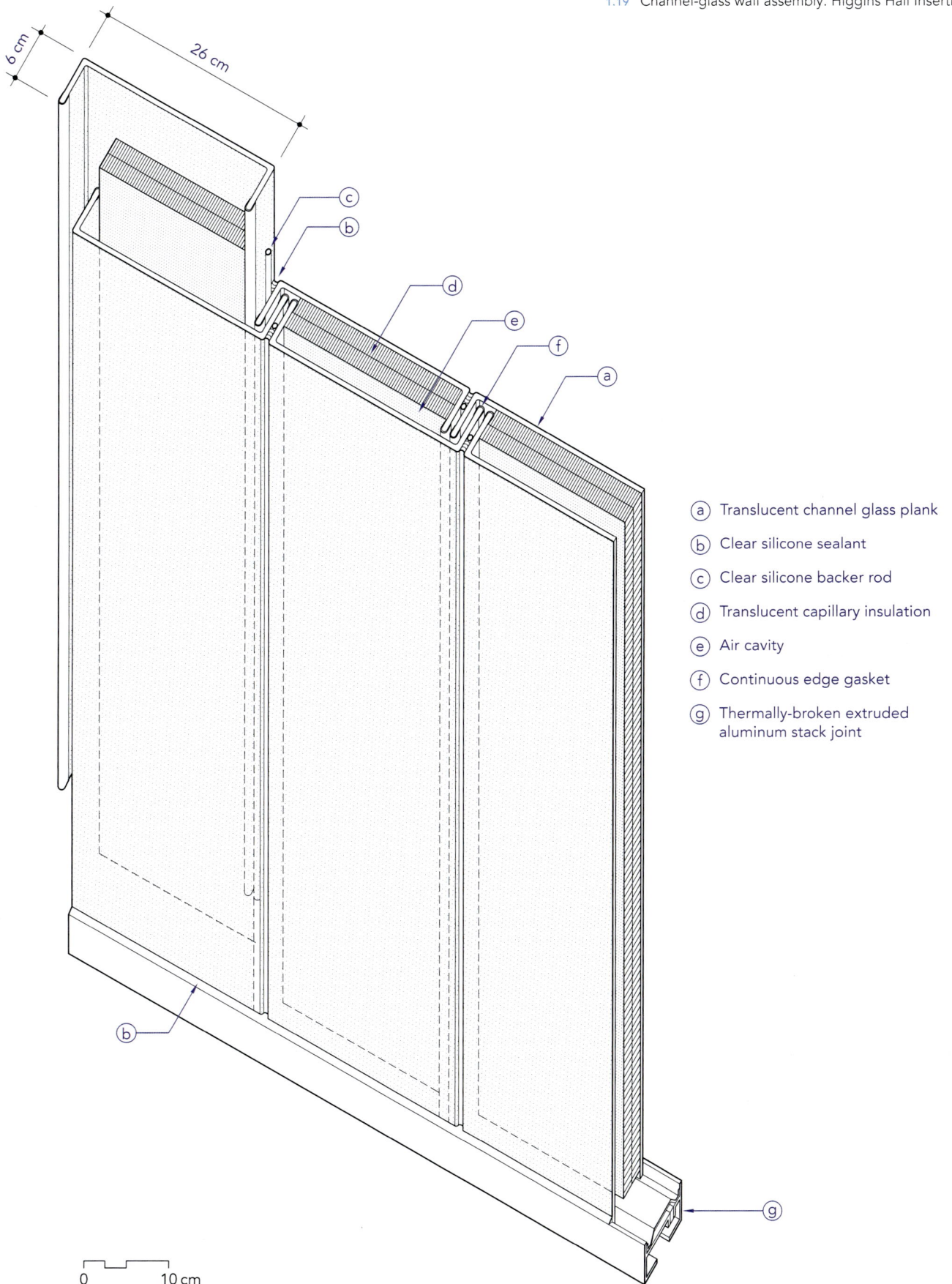

6 cm

26 cm

(a) Translucent channel glass plank

(b) Clear silicone sealant

(c) Clear silicone backer rod

(d) Translucent capillary insulation

(e) Air cavity

(f) Continuous edge gasket

(g) Thermally-broken extruded aluminum stack joint

0   10 cm

1.20 Entrance lobby. Nelson-Atkins Museum of Art, Kansas City, Missouri, USA. Steven Holl Architects, 2007.

As noted above, before being introduced to the US channel glass had been used in Europe for many years. The Neanderthal Museum (1996), designed by Kelp, Krauss, and Bruandlhuber, and built in Mettmann, Germany, includes a curvilinear three-story façade of 4-meter tall (13 feet) channel-glass planks. However, in this case the channel glass is used merely as an exterior cladding —the glass is installed over a solid, insulated concrete wall, thus imparting none of its light-diffusing qualities to the interior. The Boxtel Police Station (1997) in Boxtel, the Netherlands, by Wiel Arets Architects, uses an outer skin of single-layer channel glass to continuously wrap the building, paired with an inner skin that varies between a solid concrete wall and a second layer of channel glass, depending upon the lighting requirements of the interior spaces. Similarly to Higgins Hall, the channel-glass skin is occa- sionally interrupted by transparent glass windows for views and ventilation. Steven Holl Architects have designed channel-glass walls in various formulations for a string of recent projects follow- ing Higgins Hall. These include the Swiss Embassy Residence (2006) in Washington, D.C., the School of Art & Art History (2006)

at the University of Iowa, the Nelson-Atkins Museum of Art (2007), the Herning Museum of Contemporary Art (2009), and the Nanjing Sifang Art Museum (2012).

A profile of Steven Holl Architects prepared by the firm sum- marizes the importance of material experimentation in their work:

The materials of architecture communicate through reso- nance and dissonance, just as instruments in musical composition. Architectural transformations of natural mate- rials, such as glass, stone, or wood, produce thought and sense-provoking qualities in the experience of place.[18]

The Higgins Hall Insertion reflects this approach—it is an essay in materiality that draws upon history as much as new technological advances, and in the end its thesis resides in the unique experi- ential qualities of the built space. Like the Maison de Verre before it, the Higgins Hall Insertion offers a unique physical embodiment, or solidification, of light filtered through its innovative building skin.

1.21 Higgins Hall Insertion, Brooklyn, New York, USA. Steven Holl Architects, 2005.

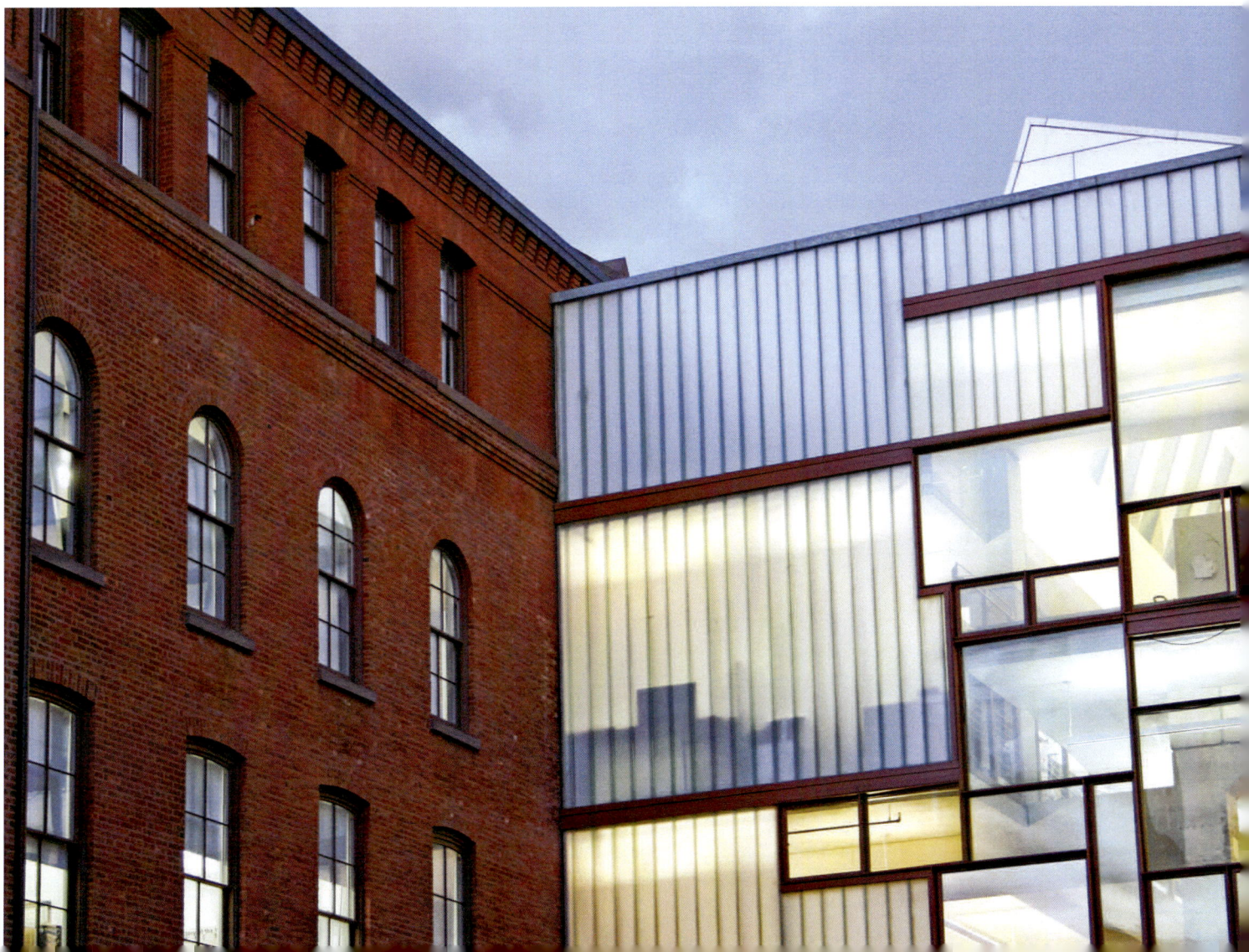

1.22 Higgins Hall Insertion, Brooklyn, New York, USA. Steven Holl Architects, 2005.

2
Art House Cinema

MUSEUM OF MODERN ART, NEW YORK CITY, USA, 1939

KUNSTHAUS BREGENZ, BREGENZ, AUSTRIA, 1997

The 1998 Mies van der Rohe Award for European Architecture was presented to the Kunsthaus Bregenz, a contemporary art museum which had opened the previous year on the eastern shore of Lake Constance in the small town of Bregenz, Austria.[1] The museum building, designed by Peter Zumthor, had been immediately hailed by the architectural press—"a triumph of intellect and imagination,"[2] according to one review. With its distinctive cubic form and the minimalist articulation of its double-skin envelope, the Kunsthaus quickly became an icon of late twentieth-century architecture, instantly recognizable by the simplicity of its building envelope: a relentlessly repetitive, shingled, translucent glass façade covers all four sides of the building. But this is a deceptive simplicity, as the façade is not, in fact, merely a simple surface; it contains depth and contradiction within its multiple layers, alternately hidden and revealed as one engages the building. And the museum's significance derives not only from its form or aesthetic appearance, but more importantly, from its unique performance—that is, how the architecture interacts with people, art, and light.[3]

In his prophetic book, *Glasarchitektur*, published in Berlin in 1914, Paul Scheerbart envisioned a future world illuminated by buildings constructed primarily of glass, writing:

> I have so often said that the double walls are there not merely to maintain the temperature of the room, but to accommodate the lighting elements . . . With this type of lighting the whole glass house becomes a big lantern which, on peaceful summer and winter nights, shines like fire-flies and glow-worms. One could easily become poetic.[4]

Scheerbart could have been describing the Kunsthaus Bregenz and its double-skin enclosure, which transforms at night into a glowing box. Scheerbart's tendency toward biological metaphor has often been echoed in critics' descriptions of Zumthor's design: "the skin has the elegance and suppleness of a lizard's"[5] and is "like a chrysalis surrounding an emerging animal."[6] Zumthor himself has written that the building's skin "looks like slightly

2.1 Kunsthaus Bregenz, Bregenz, Austria. Peter Zumthor, 1997.

28

2.2 Kunsthaus Bregenz on the shore of Lake Constance, Bregenz, Austria.

2.3 Kunsthaus Bregenz, Bregenz, Austria. Peter Zumthor, 1997.

2.4 Kunsthaus Bregenz, Bregenz, Austria. Peter Zumthor, 1997.

ruffled feathers or like a scaly structure."[7] He stresses its affinity with nature and atmosphere and the ways that such factors modulate the façade, writing that "it absorbs the changing light of the sky, the haze of the lake, it reflects light and color and gives an intimation of its inner life according to the angle of vision, the daylight and the weather."[8]

Surely at least part of the widespread interest in this relatively diminutive 2,800-square meter (30,000-square foot) museum[9] stems from its assertive embodiment of the condition of translucency. This condition defines the architect's design strategy, through the primary articulation of materials and space, and its impact on the building's use as a place for the display and experience of art, both inside and out. The building skin, which hangs from its internal frame, appears like a veil that responds to changing climatic conditions, shifting in character from nearly opaque to nearly transparent. At night, the façade becomes a building-scaled lantern, illuminated by fixtures concealed within the double-skin cavity. In this case, translucency's ethereal effects can be best understood through an examination of the building's façade system, its glass fabrication technologies and subsequent on-site assembly of its components.

The outer façade, measuring 30 meters (98 feet) in height, incorporates 712 glass panels, each measuring 1.72 by 2.93 meters (5 feet, 8 inches by 9 feet, 7 inches). Each glass panel consists of two panes of 10-millimeter thick (⅜-inch) low-iron float glass laminated together with a polyvinyl-butyral (PVB) interlayer and acid-etched on the outermost surface. The exposed acid-etch treatment is what gives the glass its distinctive translucent and matte finish, which is similar in appearance to sandblasted glass but is much finer in texture.

The glass-etching process was conducted by Fällander Glass, at the company's facility in the village of Fällanden, about 10 kilometers outside of Zurich, Switzerland. Fällander specializes in custom glass etching in two forms: full-coverage (in which the entire surface of a glass pane is uniformly etched) and motif etching (in which a pattern or image is selectively etched onto a glass surface). Fällander spent about three months producing all of the full-coverage acid-etched glass for the Kunsthaus, using a process performed by hand. The float-glass panels, which have

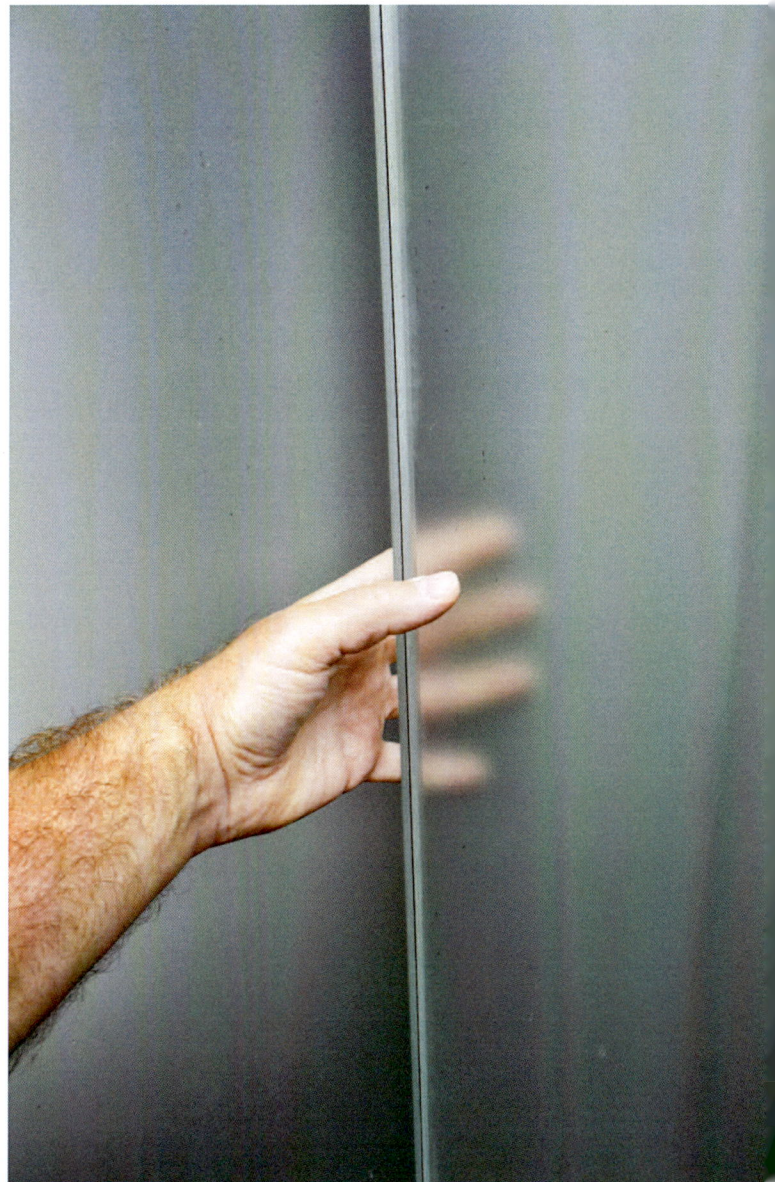

2.5 Acid-etched laminated glass. Kunsthaus Bregenz, Bregenz, Austria. Peter Zumthor, 1997.

(a) Translucent laminated low-iron glass with acid-etch on surface 1

(b) Stainless-steel support bracket and glass-stop angles

(c) Rubber gasket

(d) Steel tube, 55 × 100 mm

(e) Steel tube, 40 × 60 mm

297 cm

(b)

(c)

(d)

(e)

(a)

0   10 cm

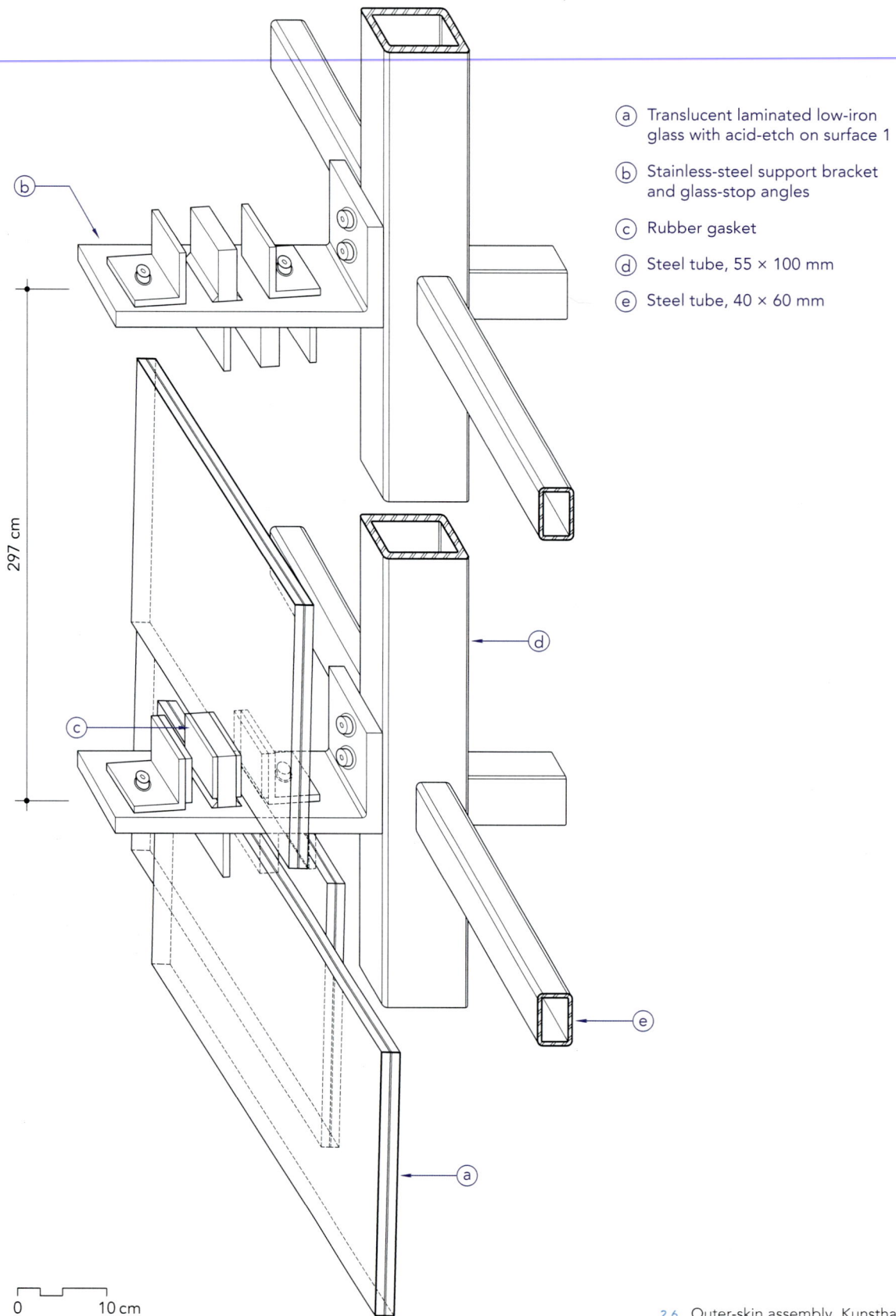

2.6 Outer-skin assembly. Kunsthaus Bregenz.

been previously cut to size and laminated by other fabricators, are first set horizontally on a work table and thoroughly cleaned. A wax border, approximately 5 centimeters (2 inches) in height is formed continuously around the edges of the glass to contain the liquid acid bath which is then poured onto the glass surface. The resulting degree of translucency, along a spectrum from nearly transparent to nearly opaque, is affected by the amount of time that the acid remains in contact with the glass—a longer duration produces more transparency, a shorter duration more opacity. For the Kunsthaus project, Fällander used their roughest (least transparent) of three standard etch levels. Following this process, the acid is drained off the glass and collected via floor grates to be reused. The glass then receives a final cleaning before being shipped to the construction site for installation.[10]

The acid-etched glass is arranged on the Kunsthaus façade in a shingle-like configuration, with each panel slightly angled, supported only at its four corners by stainless steel angles with rubber gaskets that clamp the edges of the glass and anchor it to a delicate steel-tube and tension-rod structure. This scaffold-like steel frame, as well as the second enclosure—a wall consisting of insulated concrete and glass located about 90 centimeters (3 feet) behind the glass skin—is discernible in certain lighting conditions, veiled behind the translucent glass panels. While this enigmatic, ghostly exterior appearance contributes to the iconic nature of the building, the façade contributes much more than a singular aesthetic; the outer glass skin is in fact an integral part of the lighting scheme—and thus the atmospheric character—of the interior exhibition galleries, where etched glass panels form a continuous ceiling plane illuminated by natural sunlight that has been filtered through the outer glass walls, supplemented when necessary by electric lighting concealed above the translucent ceiling.

In what may be interpreted as a rebuke of the modernist obsession with transparency, Zumthor has not yielded to the impulse to break the translucent skin and provide clear windows with views of the adjacent Lake Constance (and one can imagine the potentially spectacular views from the upper floors of the building) or views into the galleries from the exterior. In fact, the translucent skin of the building gives way to transparency in only one discrete moment: the main entrance portal's transparent glass

2.7   Stainless-steel angle supporting glass panel. Kunsthaus Bregenz, Bregenz, Austria. Peter Zumthor, 1997.

2.8   Gallery interior. Kunsthaus Bregenz, Bregenz, Austria. Peter Zumthor, 1997.

2.9   Kunsthaus Bregenz, Bregenz, Austria. Peter Zumthor, 1997.

2.10 Between the two skins. Kunsthaus Bregenz, Bregenz, Austria. Peter Zumthor, 1997.

2.11 Kunsthaus Bregenz, Bregenz, Austria. Peter Zumthor, 1997.

2.12 Kunsthaus Bregenz, Bregenz, Austria. Peter Zumthor, 1997.

2.13 Kunsthaus Bregenz, Bregenz, Austria. Peter Zumthor, 1997.

doors, which face toward the plaza, away from the lake. This stubbornness reinforces the building's overriding inward-focus toward its galleries and the art displayed there, and it contributes to the serene, introspective atmosphere that defines the gallery spaces. The relationship between the exterior skin and the interior spatial experience remains indirect, though powerful.

Despite this internal focus, the Kunsthaus Bregenz has been successful at exhibiting art not only inside its galleries but on its exterior surfaces as well. Not surprisingly, the building's translucent envelope has proven to be a worthy and inspirational venue for temporary, site-specific public art installations that use light as a primary material.[11] The cavity between the façade's two skins provides space for mounting and concealing light fixtures, while the translucent outer skin serves as a scrim to capture and disperse light, transforming the normally mute and tranquil walls into unusual conveyors of imagery and information. The location of the Kunsthaus allows viewing of such installations from the immediately adjacent sidewalk and plaza as well as from points much farther away along the lakeshore.

The museum's opening exhibition, in the summer of 1997, featured an installation by the artist James Turrell, who placed light fixtures within the façade cavity to transform the entire building into a glowing block of colored light on all four sides, shifting in hue from shades of blue and white to red and violet—abstract illuminated signage announcing the opening of the new museum. In 2001, artist Tony Oursler animated the lakeside glass façade, using it as an immense projection screen, 23 by 23 meters (75 by 75 feet), for his video and sound installation, "Flucht," consisting of moving images of human faces (projected from specially erected scaffold towers) and their voices (transmitted from loudspeakers within the façade). Jenny Holzer's 2004 installation, "Xenon for Bregenz," featured a spectacle of large-scale projected text related to the historical involvement of the USA in Iraq, injecting a timely and politically provocative message into the public realm. The Kunsthaus façade, with its uncommon material deployment and structural arrangement, has served as a neutral infrastructure for art, enabling these artists, and many others, to realize diverse large-scale applications for their experiments with

2.14 Kunsthaus Bregenz, Bregenz, Austria. Peter Zumthor, 1997.

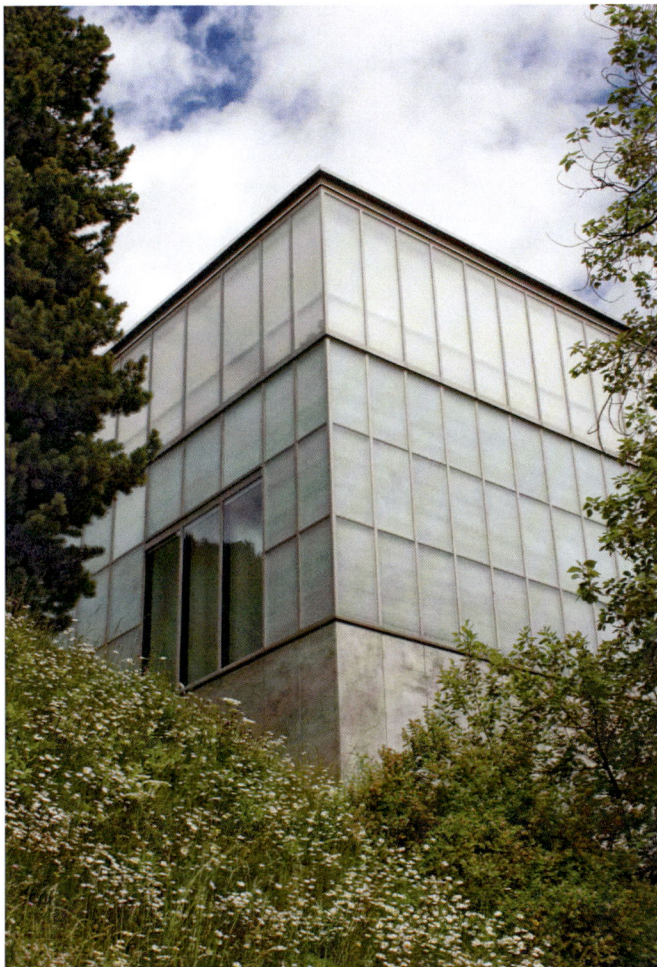

2.15 Kirchner Museum, Davos, Switzerland. Gigon and Guyer, 1992.

2.16 Kirchner Museum, Davos, Switzerland. Gigon and Guyer, 1992.

light, and it therefore stands as a unique example of the potential federation of architecture and public art.

Kunsthaus Bregenz is perhaps the best known representative from a group of museum facilities that convey a primary concern for light and translucency. One such building is the Kirchner Museum in Davos, Switzerland, which opened in 1992. Designed by Annette Gigon and Mike Guyer, this single-story museum houses a collection of works by the Swiss artist Ernst Ludwig Kirchner. The gallery spaces utilize a system of suspended translucent glass ceilings that can be illuminated electrically or naturally via a continuous translucent glass clerestory that spans from ceiling to roof. The sectional arrangement of these components is similar conceptually (though quite different visually) to Zumthor's double-skin and glass-ceiling design in Bregenz. Perhaps not incidentally, the Kirchner Museum's acid-etched glass panels were fabricated by Fällander Glass, the same company which later supplied glass for the Kunsthaus Bregenz.

The influence of these two buildings may be seen in several later museum projects that make translucency a defining charac-

teristic. The Auguste Rodin Museum (1999) in Seoul, Korea, was designed by KPF Associates to permanently house two Rodin sculptures, *The Gates of Hell* and *The Burghers of Calais*. The museum's ceiling/roof assembly and its sinuous double-skin walls are formed by translucent glass panels, point-supported from a steel structure housed within the wall cavity. The entire enclosure system thus becomes a source of evenly dispersed light throughout the pavilion-like gallery. The shingled glass panels that form the cone-shaped entrance structure at Gluckman Mayner Architects' Mori Arts Center (2003) in Tokyo are printed with a silkscreened pattern of ceramic frit dots which render the glass somewhat translucent while still maintaining a high degree of visible-light transmission. The Figge Art Museum (2005), in Davenport, Iowa, by David Chipperfield, presents a striking exterior wrapped entirely by a translucent double-skin glass enclosure printed with a white ceramic-frit line pattern; however, except for a few discrete locations, the glass skin does not contribute much to the experience of the interiors, which are for the most part generic, artificially lit gallery spaces.

2.17   Auguste Rodin Museum, Seoul, Korea. KPF Associates, 1999.

2.18   Mori Arts Center, Tokyo, Japan. Gluckman Mayner Architects, 2003.

2.19   Figge Art Museum, Davenport, Iowa, USA. David Chipperfield, 2005.

Steven Holl Architects' addition to the Nelson-Atkins Museum of Art (2007) in Kansas City, Missouri, exhibits masterful manipulation of light, using sandblasted channel-glass planks to form a series of linked translucent gallery "lenses" sited within a sprawling sculpture garden.[12] In a similarly poetic and earthbound vein, Terry Pawson Architects' Visual Centre for Contemporary Art (2009) in Carlow, Ireland, is an assemblage of rectilinear volumes of various height wrapped by a double-skin system incorporating an outer layer of translucent glass that selectively transmits diffuse natural light to interior gallery spaces and glows from within at night. Each of these buildings—though varied in size, location, and success—shares a common precedent which, decades earlier, introduced the effects of translucency to modern museum architecture in a highly visible location: the original 1939 Museum of Modern Art in New York City.

* * *

The Museum of Modern Art (MoMA), an early champion of modernism in America, is a cultural institution whose mission includes "to provide a detailed but clearly intelligible history of modern art" and "to engage the public with its programs and ideas."[13] Since the construction of its original building on 53rd

2.20 Nelson-Atkins Museum of Art, Kansas City, Missouri, USA. Steven Holl Architects, 2007.

Street between Fifth and Sixth Avenues in midtown Manhattan in 1939, MoMA has purposefully pursued engagement with the public not only through its collections and exhibitions but also through its architecture, which can be viewed as the public projection of the museum's identity and has itself become a historical record documenting attitudinal shifts related to modernity. Through a series of major expansions and renovations conducted over the course of several decades, the museum has conceived its own architecture as a way to directly engage the public and to express contemporary interpretations of the modern condition.

Following its most recent expansion, which was designed by the Japanese architect Yoshio Taniguchi in association with the New York firm KPF and completed in 2004 at a cost of 425 million dollars, MoMA now occupies a massive complex encompassing more than 59,000 square meters (640,000 square feet) of space in a series of linked buildings spanning between 53rd and 54th

Streets. This aggregation of buildings can be read as a physical history of the museum's evolution over its 80-year existence. MoMA has conducted six major building expansions since construction of its original 1939 building: the Grace Rainey Rogers Annex (1949–51) by Philip Johnson, the Abby Aldrich Rockefeller Sculpture Garden (1952–53) by Johnson, the integration into MoMA of the former Whitney Museum building (1963–68) by Johnson and Augustus Noel, the East Wing addition (1964) by Johnson, the Museum Tower Wing (1979–84) by Cesar Pelli, and the latest expansion (1997–2004) by Yoshio Taniguchi and KPF. Taniguchi's project included a major addition that dramatically reconfigured the internal spaces of the museum as well as the renovation and restoration of certain elements of earlier buildings within the complex. Important among these is the restoration of MoMA's original 53rd Street façade to its 1939 configuration, including one of its unique and defining components: its translucent Thermolux curtain wall.

Despite the historically close association of MoMA's identity with its architecture, for the first decade of its existence the museum lacked a permanent purpose-built home. MoMA was founded in 1929 by seven prominent art collectors with the aim of increasing public access to contemporary art through temporary exhibitions of painting, sculpture, and, later, industrial arts, photography, film, and architecture.[14] For the first two years of its existence, MoMA held its exhibitions in rented space on the twelfth floor of the Heckscher Building on Fifth Avenue. As early as 1930, however, museum trustees were making preliminary plans for constructing a new building specifically for MoMA. The young firm of Howe & Lescaze (soon to become famous for their Philadelphia Saving Fund Society Building) was engaged to produce a fascinating series of design schemes, ultimately unbuilt, for a new museum on a hypothetical site in Manhattan. One of these, known as Scheme Four, is notable for a system of translucent clerestories designed to introduce diffuse natural light into the galleries through a glazed ceiling, similar to systems deployed decades later at Gigon and Guyer's Kirchner Museum and Zumthor's Kunsthaus.[15]

The Howe & Lescaze plans were put on hold indefinitely in 1932 when John D. Rockefeller, Jr., donated a four-story town-

2.21 Museum of Modern Art, New York City, USA. Goodwin and Stone, 1939.

THE MUSEUM OF MODERN ART

ART IN OUR TIME

2.22 Construction photo, 12 September 1938, Museum of Modern Art, New York City, USA. Goodwin and Stone.

house at 11 West 53rd Street that would become the museum's home for the next six years (and would eventually be demolished, along with several adjacent structures, for construction of the new building). As a further testament to the museum's growing interest in new architecture, MoMA's last exhibition at its old space in the Heckscher Building was also its first exhibition of architecture: the highly influential International Style exhibition of 1932, organized by Henry-Russell Hitchcock and Philip Johnson.[16] Shortly after this exhibition, Johnson was named the first chairman of the museum's newly formed Architecture Department. By 1935, the popular museum had outgrown its townhouse, and the process of creating a new permanent home for the museum began in earnest. By that time, the partnership of Howe & Lescaze had been dissolved, and the commission for the new building went to the architect Philip Goodwin, a museum trustee, and his junior partner, Edward Durell Stone. Although supported by the museum's Building Committee, the relatively conservative Goodwin and Stone were not the first choice of MoMA Director Alfred H. Barr, Jr., who had wanted Mies van der Rohe to be the architect and had also considered Walter Gropius and J.P. Oud. Having been overruled, a

frustrated Barr resigned from the Building Committee in 1936 (but continued as Museum Director).[17]

The museum was built at a cost of two million dollars and opened in May of 1939. Goodwin and Stone's design incorporated six levels: a ground-floor entrance and lobby, exhibition galleries on the second and third floors, administrative office space on the fourth and fifth, and a sixth-floor penthouse and terrace for museum members and trustees. The main, south-facing 53rd-Street façade consisted of transparent glass at the ground level, with the upper floors clad in white marble panels surrounding a two-story expanse of steel-framed translucent glass curtain wall at the gallery levels and two horizontal bands of steel-sash windows with clear glass extending the full length of the building at the office floors. The museum's stark geometric façade, built out to the lot line, contrasted sharply, and intentionally so, with the adjacent row of traditional four-story brownstone townhouses set back along 53rd Street. This façade represents one of the earliest examples of International Style architecture on an institutional scale in the USA and would become an emblem of modernity in America. The critic Paul Goldberger has written that

2.23 Museum of Modern Art, New York City, USA. Goodwin and Stone, 1939.

the design "elucidated superbly the 'International Style' principles: tensile, sharp, crisp, it remains one of New York's great early modern monuments."[18]

The largest area of the main façade, and its most dominant and defining feature, is the double-height translucent Thermolux curtain wall enclosing the primary exhibition galleries and an open staircase at the second and third floors. Here, translucency is a particularly intriguing characteristic of the design, especially given the widespread emphasis on transparency as an ideal in the modern architecture of the early twentieth century. In contrast to the iconic 1920s and 1930s architecture of Alfred Barr's preferred designers—Mies van der Rohe and Walter Gropius, who had each embraced the ideal of transparency in both their built and theoretical works—Goodwin and Stone's design for MoMA relegated transparent glass to the utilitarian functions of entry/lobby and office. The heart of the museum experience—its art galleries—was clad in the ghostly, ambiguous translucency of Thermolux.

Thermolux, a glass product fabricated by the Libbey-Owens-

Ford Company, consisted of a thin, white, translucent glass-fiber "tissue" laminated between two panes of clear glass. At close range, the striations of individual glass fibers could be seen through the clear glass casing. It was a hand-assembled product, and therefore the array of fibers in each panel would appear as a slightly different grain. Thermolux transmitted diffuse natural sunlight into the gallery spaces by day and was illuminated like a lantern by internal lighting at night. Although the glass precluded direct vision, shadows of passing clouds would register subtly on the interior side of the tall, white, glass walls by day; slight reflections of sunlight off passing vehicles would flash briefly across the inside glass surface, providing at once an abstract connection to and removal from the outside world of the city. With interior lighting at night, shadows projected by Alexander Calder's mobile, Lobster Trap and Fish Tail (commissioned to hang over the main stair just inside the Thermolux wall) could be seen on the exterior façade. The delicate texture of the Thermolux glass fibers also complemented the natural veining of the façade's stone cladding,

2.24 Interior staircase and Thermolux curtain wall. Museum of Modern Art, New York City, USA. Goodwin and Stone, 1939.

which consisted of 100-millimeter thick (4-inch) white Georgia Marble panels. According to the manufacturer of Thermolux, "the tiny fibers of translucent glass, each about six ten-thousandths of an inch in diameter, break up light rays into thousands of miniature rays, giving a soft, evenly diffused light over a wide area."[19]

Goodwin and Stone's construction details indicate that 155 panels of Thermolux glass, each measuring approximately 1.2 by 1.5 meters (4 by 5 feet), were supported by a framework of continuous T-shaped steel mullions running horizontally and vertically, which were anchored to the concrete floor slab at each level and clad on the exterior with stainless steel caps. It is interesting to note that in Goodwin and Stone's earliest façade designs, the gallery spaces were essentially windowless and dark, clad entirely with opaque marble-faced walls. Thermolux came onto the market just in time to be incorporated into the project and new elevations featuring the Thermolux façade were finalized just months before construction began.[20]

By the time of Cesar Pelli's major alterations and additions to the MoMA complex in the 1980s, the 53rd Street façade as originally designed and constructed—including the Thermolux curtain wall—would unfortunately exist only in photographs and drawings. Eventually, as the museum's collection grew and additional space was needed for hanging artwork, a plaster partition wall was built just inside the curtain wall, completely hiding the Thermolux glass from view. Eventually all of the Thermolux glass panes were removed and replaced with spandrel glass (essentially clear glass that was painted white on the back surface, rendering it opaque and resulting in a dull, uniform appearance). This may have been partly due to the unavailability of replacement glass, as Libbey-Owens-Ford had discontinued production of the Thermolux product at some point in the late 1940s. At the same time as the glass replacement, the steel mullion system was removed and replaced with a standard extruded-aluminum system.[21] The magical quality of light transmission, an important part of experiencing the building and its art, and the distinctive textural quality of the

122 cm

152 cm

(b) Steel T mullion, 64 × 100 mm

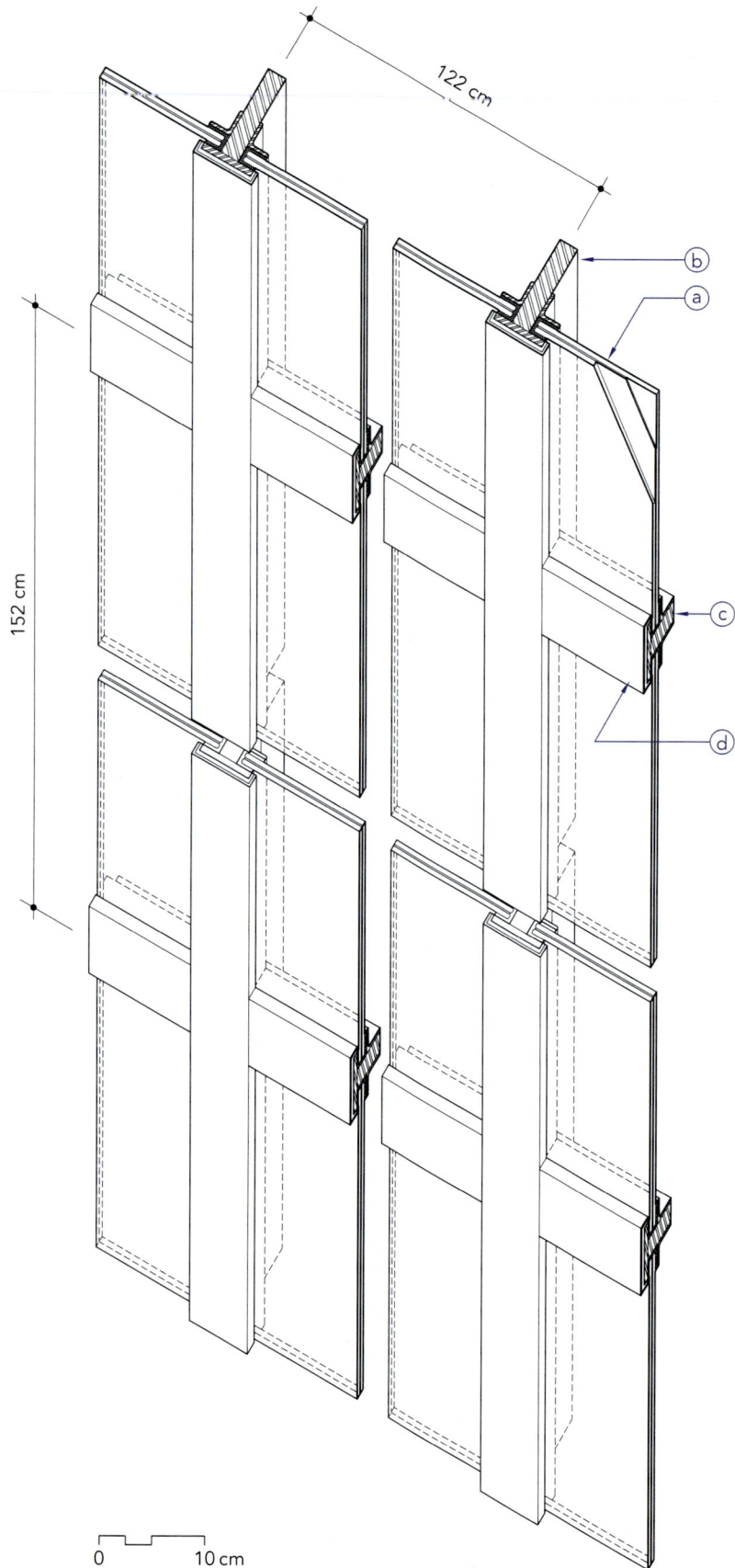

(a) Laminated glass with translucent Thermolux interlayer

(c) Steel T mullion, 64 × 60 mm

(d) Formed stainless steel cladding

(a) Laminated glass with translucent Thermolux interlayer

(b) Steel T mullion, 64 × 100 mm

(c) Steel T mullion, 64 × 60 mm

(d) Formed stainless steel cladding

0   10 cm

2.25   Thermolux curtain wall. Museum of Modern Art.

Thermolux glass fibers were thus lost. Visitors to the museum after this point would see no indication of the translucent "great window," which the critic Lewis Mumford had hailed as one of the building's "specially rewarding spots" in his 1939 review of the new building in *The New Yorker*.[22]

Goodwin and Stone's original design was fortunately revived, however, during MoMA's latest expansion. Taniguchi's 2004 project added entirely new wings to the museum complex, creating a much larger presence in midtown Manhattan by virtue of its immense size, its pass-through lobby with entrances on both 53rd and 54th Streets, and the distinctive design of its white glass and black granite cladding. But part of the project also addressed the original 1939 building façade. MoMA and its design team, which included Taniguchi, KPF, and the façade consultants of R.A. Heintges & Associates, would attempt to "un-do" this history of piecemeal transformation and return the façade to its original design by rebuilding those elements which had been lost over time, including the Thermolux curtain wall and the curvilinear entrance canopy and storefront which had been demolished during Philip Johnson's 1950s renovations.[23] The design team was successful in locating a Swiss company that was currently still producing a version of Thermolux glass (its product, called Termolux, consisted of a nearly identical assembly of glass and fibers). Newly revealed inside and out, the restored 2004 Thermolux curtain wall, translucent and luminous like its predecessor, creates a unique experience for the museum visitor that has not been possible for decades, and reconnects the contemporary incarnation of MoMA to its past.

With the design and construction of its first facility in 1939, MoMA sought "to produce a building that would serve as a three-dimensional demonstration of all that the Museum stood for, commensurate with a reputation grown to nationwide dimensions."[24] The historian Talbot Hamlin wrote of the newly opened museum:

> Since architecture is one of the arts in which it is deeply interested, its own new building had to serve as a public

2.26 Restored Thermolux curtain wall, interior, 2004. Museum of Modern Art.

2.27 Restored Thermolux curtain wall, interior, 2004. Museum of Modern Art.

2.28 Restored façade with Thermolux curtain wall, 2004. Museum of Modern Art.

evidence of its aims and ideals. Thus, the design . . . became, in all truth, a part of the museum collection—the only part permanently and indefinitely on display.[25]

MoMA, the Kunsthaus, and other projects discussed in this chapter illustrate the ways that translucency has been deployed within the art museum typology to create cultural identity and to explore the phenomenon of projection, both literally, as in the projection of images upon the Kunsthaus façade, and symbolically, as in the projection of a concept of modernity at MoMA.

3
Crystallization

3.1 Steiff Factory Building, Giengen an der Brenz, Germany. Richard Steiff, 1903.

Due to its inherent relationship with the perception of manipulated light, the condition of translucency in architecture is often associated with primarily subjective aims: the creation of spectacle, affect, or atmosphere. But translucency also has the potential to address practical issues of function and technical performance in buildings of certain usage which may require, for instance, specific lighting conditions or degrees of privacy or publicity. The incorporation of translucent materials in the design of a building skin may therefore be related to *program* as much as (if not to the exclusion of) spectacle or aesthetic expression, phenomena which become by-products of the main intent. In these types of buildings, which are the focus of this chapter, the enclosure system may

be read as an embodiment, or crystallization, of rational programmatic concerns, reflecting the fundamental importance of light in the activities taking place within and often accompanied by a prioritization of materiality and performance over form.

Among the many influential industrial buildings that contributed to the rise of modern architecture in the early twentieth century, the lesser-known Steiff Factory Building of 1903 stands out for its stunningly inventive translucent double-skin glass enclosure. This three-story loft structure, which still stands today in Giengen, Germany, was built by the Steiff Company, a manufacturer of dolls and other toys, for purely utilitarian purposes: to provide flexible spaces for manufacturing with abundant natural

3.2 Aerial view of the Steiff Factory complex, 1910.

light and a comfortable interior environment. The first of these requirements led to the design of exterior curtain walls made largely of glass. The second led to the invention of a double-skin envelope system, an ingeniously artful response to the need for thermal insulation and one of the earliest examples of a construction methodology that continues to be investigated and implemented a full century later. Furthermore, as Steiff was facing quickly growing demand for its famous teddy bear, the factory had to be built inexpensively and fast, which also factored into the selection of materials and fabrication techniques.

The design of the building is most often attributed to the engineer Richard Steiff, who helped manage the company and was the nephew of its founder, Margarete Steiff. With its steel-framed structure and continuous glass skin on all sides, the Steiff Factory was unlike any other building in the small village of Giengen or elsewhere, for that matter.[1] It has been speculated that Richard Steiff's design may have been influenced by a trip to England, where he likely experienced Joseph Paxton's Crystal Palace, and also by his father Friedrich Steiff, a builder who had visited Chicago for the 1893 World's Columbian Exhibition, where the elder Steiff would have been exposed to the city's growing collection of innovative steel-framed skyscrapers with extensive glazing.[2] In its tectonics, Steiff's design also recalls the iron-framed, glass-clad greenhouses of the mid- to late

Extension of the Steiff Factory, 1904.

nineteenth century, including the Palm House (1859, demolished 1910) at the Royal Botanical Garden in Berlin-Schöneberg and the Conservatory (1861) at the Botanical Garden in Breslau.[3]

The Steiff Factory Building (also known as the East Building within the Steiff complex) is a simple rectangular volume, covering an area of 12 by 30 meters (40 by 98 feet) and standing 9.4 meters tall (31 feet) with a shallow-pitched shed roof. The immediate success of the building, and of the Steiff Company and its dolls, led quickly to the construction of two similar but much larger structures adjacent to the original. In 1904 a carpenters' shop was built (known as the South Building), followed by an additional factory structure in 1908 (the West Building). The following analysis will

examine the original 1903 building and the construction methods initiated there, which formed the model upon which the later buildings were based.[4]

The building skin of the Steiff Factory is a true curtain wall—it is a non-load-bearing enclosure separated from, yet anchored to, the building's structural steel frame.[5] The structural plan consists of five bays lengthwise and three across, framed by riveted steel beams and columns bearing on a concrete base. Diagonal bracing for lateral wind resistance is also incorporated into the walls and floors. This prefabricated steel skeleton was erected in just a few days by the firm of Eisenwerk München.[6] On each of its four elevations, the building's envelope is formed by walls consisting

3.4 Construction of the Steiff Factory Building, 1903.

of two layers of 3-millimeter thick (⅛ inch) single-pane cathedral glass supported by a simple grid of T-shaped rolled steel mullions. Each pane measures approximately 60 by 90 centimeters (2 by 3 feet). The T-shaped mullions are each 25 millimeters wide (1 inch) by 35 millimeters deep (1-⅜ inch). The two glass walls are separated by a space of about 25 centimeters (10 inches), creating an air cavity that serves as a thermal buffer between internal and external environmental conditions.[7] The structural steel columns are also located within this interstitial space. The inner glass wall spans from the top of each floor slab to the ceiling above, while the outer glass wall spans continuously from the building's base to the roofline. Box windows, which can be manually opened or closed, are placed intermittently on each façade and span

between the two glass walls to link interior with exterior. As the building lacked mechanical air-conditioning, during the warm summer months the interior space was to be kept cool through natural cross-ventilation enabled by these windows, in conjunction with shading provided by adjustable interior blinds. Eventually these methods proved inadequate against the greenhouse effect, and mechanical ventilator fans were later installed in strategic locations along the curtain walls. During cold winter months, the air cavity between the glass walls helps maintain thermal separation between inside and out, but is supplemented by a low-pressure steam heating system.

The panes of cathedral glass, which were typically made by rolling molten glass into sheet form, are technically transparent

90 cm

(a) Translucent cathedral glass

(b) Steel T mullion

(c) Steel angle sill

(d) Glazing compound

(e) Floor slab

(f) Wood flooring

(g) Structural steel-channel beam

0   10 cm

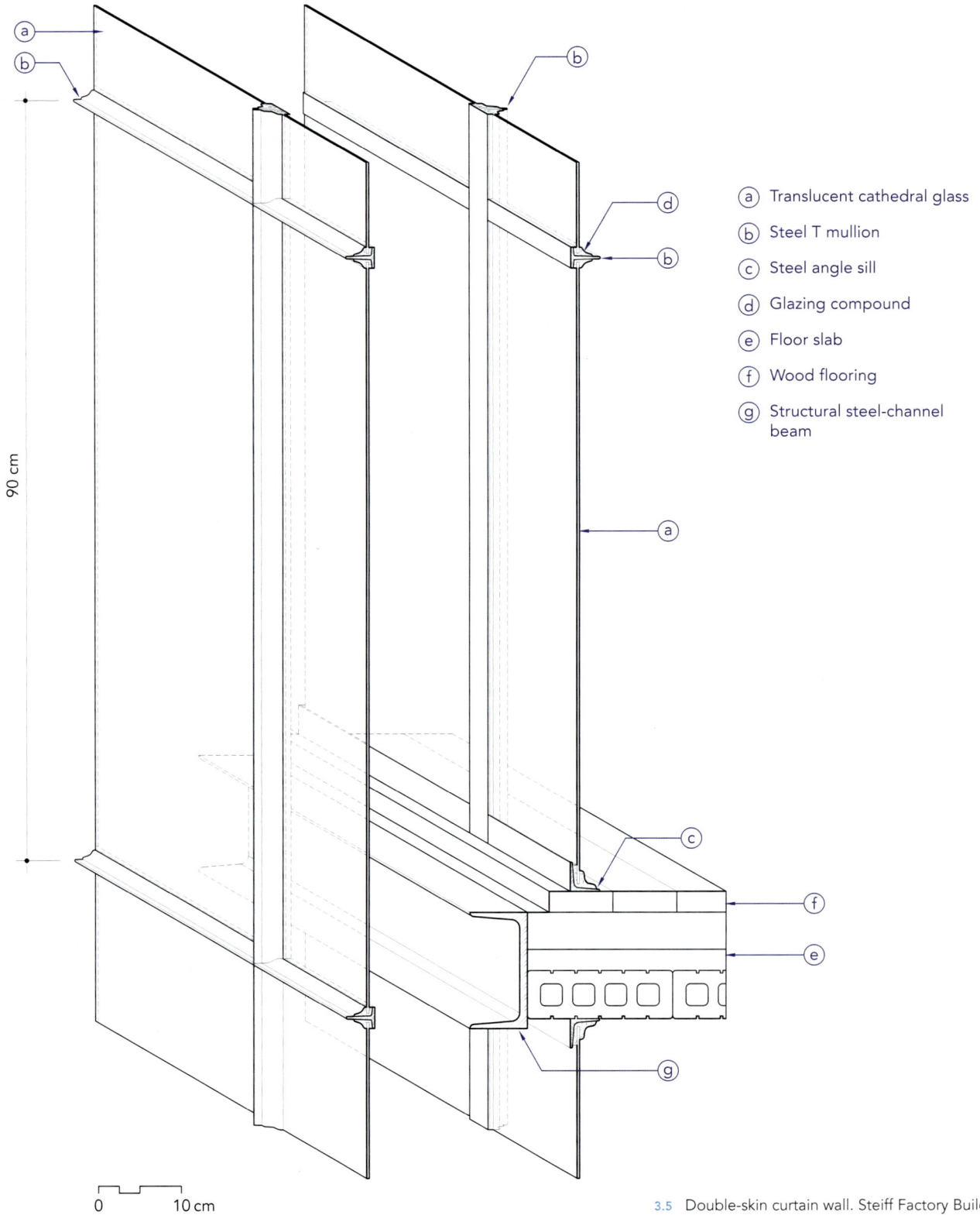

3.5   Double-skin curtain wall. Steiff Factory Building.

3.6 Interior, Steiff Factory Building, Giengen an der Brenz, Germany. Richard Steiff, 1903.

but have a rippled texture on one surface. This texture renders the glass translucent in appearance, somewhat obscuring vision and preventing glare caused by excessive direct sunlight. It also provides diffusion of natural light deeper into the interior space than would smooth-surfaced glass—of significant advantage in a workshop space. It is unclear whether Richard Steiff's specification of this particular glass type was based mainly on its light-projecting capability or on the fact that it was a less expensive alternative to smooth transparent glass. Nevertheless, the use of rippled glass lends the building its characteristically hazy translucency, revealing the structural elements immediately behind the glass but offering only shadowy hints of the people and objects further away.

Considering its innovations in materiality and construction, the Steiff Factory Building was clearly ahead of its time. It should be noted, however, that for many decades the building was essentially excluded from the history of modern architecture, perhaps because it was not designed by an architect and was not part of a larger body of work.[8] The Steiff Building has been overshadowed

by other German industrial buildings that closely followed it, including the famous AEG Turbine Factory and the Fagus Shoe-Last Factory. Though both these buildings exhibit finely detailed single-skin curtain walls, neither reaches the level of sheer invention achieved at the Steiff Factory, which today would be considered the most modern of the three. Designed by Peter Behrens and built in Berlin in 1909, the AEG Turbine Factory balances its vertical bands of curtain wall with exposed steel columns and massive corner piers. Its segmented pediment and columnar expression seem to allude to the temples of ancient Greece.[9] Two years later, the three-story Fagus Shoe-Last Factory, designed by Walter Gropius and Adolph Meyer, was built in Alfeld an der Leine. Here the steel-framed curtain wall is still confined to the space between columns (rather than being a continuous surface outboard of the columns, as at Steiff), although it is uninterrupted as it extends vertically beyond the edge of each floor slab and as it wraps around column-free corners. Another building by Gropius and Meyer, their Model Factory Building constructed for the 1914 Werkbund Exhibition in Cologne, illustrates their

3.7 Steiff Factory Building, Giengen an der Brenz, Germany. Richard Steiff, 1903.

evolving concept of the curtain wall and reaches incrementally closer to the ideal of the continuous glass skin as already successfully achieved in Giengen a decade before. This progression would reach its apex at the Bauhaus Building of 1926 in Dessau, where Gropius and Meyer designed a multi-story continuous glass curtain wall that has become an icon of twentieth-century architecture. In 1958, Ove Arup and Partners designed a factory building in England that even more closely embodies the strategies of materiality and construction originated by the Steiff Factory: the CIBA Plant and Laboratories in Duxford. This four-story structure, built for the production of epoxy resins, is a simple cubic volume with a very shallow pitched roof. The building is clad on all four sides with light-diffusing translucent glass in a steel-framed curtain-wall system. Though only a single rather than double skin, the curtain wall is suspended in front of the building structure in a configuration similar to Steiff's. In its material expression and proportions, the CIBA Plant suggests a midcentury "missing link" between the Steiff Factory and Peter Zumthor's 1997 Kunsthaus Bregenz (see Chapter 2).

Though surpassed in recognition by many of these later buildings, the Stieff Factory stands as a pioneering work of architecture that, in retrospect, appears unflinchingly modern (even as it eschews the modern tenet of transparency). Through addressing utilitarian concerns for the maximization of light, the movement of air, and overall environmental performance, it established the main principles of double-skin curtain-wall technology that continue to inform today's experimental façade designs, and, as a by-product, created a unique visual identity and aesthetic expression which also resonate deeply in the contemporary scene.

* * *

Like the Steiff Factory built a century before, the Laban Dance Center is an inwardly focused container of activities which require large, open, flexible spaces that benefit from an abundance of diffuse natural light. These activities include the teaching, practice, and performance of contemporary dance. Also, like the Steiff Factory, the Laban Dance Center utilizes an inventive, translucent,

3.8 CIBA Plant and Laboratories, Duxford, England. Ove Arup and Partners, 1958.

3.10 Laban Dance Center, Deptford, London, UK. Herzog & de Meuron, 2003.

double-skin envelope that is unique in its materiality. More so than the Steiff Factory, however, Laban intentionally blurs the boundaries between pragmatic obligations and art.

Named for the choreographer and modern-dance pioneer Rudolf Laban, the building was designed by Swiss architects Herzog & de Meuron and constructed in 2003 on a two-acre site adjacent to Deptford Creek in an industrial district in southeast London.[10] With a total area of 7,800 square meters (84,000 square feet) on three levels, Laban claims to be the largest purpose-built modern dance center in the world. It has also become a cultural landmark, winning the 2003 Stirling Prize from the Royal Institute of British Architects.[11]

From a distance, the building takes the general form of a simple industrial warehouse; at closer inspection, however, it becomes clear that there is more at work here. The building form is basically rectangular in plan, with the exception of the gentle curve of one façade that contains the main entrance and faces a landscaped forecourt with sculptural earthen mounds.[12] A large central theater is surrounded by a series of 13 dance studios, each of a different size, height and color scheme. The building is con-

structed with a cast-in-place concrete frame and is wrapped on all sides by a double-skin cladding system incorporating an outer layer of translucent polycarbonate panels, punctuated by the occasional reflective-glass window, and an inner wall of translucent float glass. The potential impulse to simply place dance on display through the design of a transparent glass box, for instance, has been suppressed and supplanted by a desire to provide optimum interior conditions for dance, achieved through engagement with concepts of depth and layering. The selection of polycarbonate as the main exterior façade material—an unusual choice more commonly associated with garden sheds or cheap industrial buildings, rather than cultural institutions—is an inspired move and one that is consistent with the modus operandi of the building's designers.

In an analysis of the architects Herzog & de Meuron, Rafael Moneo writes:

The importance given to matter and materials in their work —with the attendant importance of these in construction, through technique—comes with a deliberate suppression

3.11 Laban Dance Center, Deptford, London, UK. Herzog & de Meuron, 2003.

of image, a conscious abandonment of iconographic references. . . . We could therefore say that Herzog & de Meuron ignore iconography, renounce expression and communication, in order to win back for architecture the gravity of construction and bring about a joyful rediscovery of the fundamental nature of materials.[13]

In nearly every project by Herzog & de Meuron there is a strong component of material research and experimentation. They often explore uncommon applications for common building materials. At their Dominus Winery (1998) in California, for instance, the architects formed walls with stone-filled gabions (usually used as retaining walls in road construction) resulting in an elegant light-filtering device. At the St. Jakob Stadium (2002) in Basel they repurposed ordinary plastic dome skylights as wall cladding, creating an undulating surface that glows at night. For the gates at 40 Bond Street (2007) in New York City, they formed cast aluminum into a three-dimensional representation of graffiti tags. Achieving the effects of translucency through material

manipulation has been another recurring theme in their work: see, for example, the sandblasted glass clerestories at the Goetz Collection Gallery (1992) in Munich, the two-story translucent glass penthouse (known as "the lightbeam") atop the Tate Modern Gallery (2000) in London, the illuminated ETFE cushions at the Allianz Arena (2005) in Germany, and the perforated and embossed copper cladding at the De Young Museum (2005) in San Francisco. Herzog & de Meuron also often pursue collaboration with contemporary visual artists. At the Eberswalde Technical School Library (1999) in Germany, they worked with artist Thomas Ruff to develop a "tattooed" building skin through a process of silkscreening repetitive photographic images onto the façade's glass panels and stamping images into concrete panels. Herzog & de Meuron collaborated with artist Ai Weiwei on the design of the steel "bird's nest" of the National Olympic Stadium (2007) in Beijing. The Laban Center represents a convergence of these ideas: the architects collaborated with artist Michael Craig-Martin to design a translucent building skin selectively infused with subtle color by utilizing the unusual material of polycarbonate.

3.12 Tate Modern Gallery, London, UK. Herzog & de Meuron, 2000.

3.13 De Young Museum, San Francisco, California, USA. Herzog & de Meuron, 2005.

Polycarbonate is a dimensionally stable thermoplastic polymer that is relatively lightweight while also being quite stiff, durable, and resistant to impact and extreme temperatures (though less so to abrasion). It is typically formed into either corrugated sheet or hollow multi-wall panels consisting of two polycarbonate face sheets joined by an internal egg-crate grid of the same material. Polycarbonate panels are manufactured in standard widths ranging from 50 to 200 centimeters (20 to 80 inches) and in lengths up to 15 meters (50 feet) or more, a characteristic that allows the panels at Laban to run the full height of the building with no intermediate horizontal joints. Polycarbonate panels can be produced in clear and tinted formulations, and their light transmission properties range from transparent to opaque, but are most often translucent. Compared to more traditional cladding materials like glass, metal, or masonry, polycarbonate tends to be significantly less expensive. Despite this fact and its associated reputation for insubstantiality, the material has been deployed successfully in high-profile architectural projects. Bernard Tschumi Architects used curved polycarbonate panels supported by a structure of timber ribs to create the translucent shell of the Limoges Concert Hall (2007) in central France. Rem Koolhaas designed walls of translucent polycarbonate panels to enclose his

3.14 Tinted polycarbonate panels design in collaboration with the artist Michael Craig-Martin. Laban Dance Center, Deptford, London, UK. Herzog & de Meuron, 2003.

3.15 Limoges Concert Hall, Limoges, France. Bernard Tschumi Architects, 2007.

Serpentine Gallery Pavilion (2006) in London. Koen van Velsen wrapped the exterior of the Pathé Multiplex Cinema (1995) in Rotterdam with corrugated polycarbonate sheets. Herzog & de Meuron have also used polycarbonate in earlier projects, most dramatically at the Ricola Production and Storage Building (1993) in Mulhouse, France, where the polycarbonate cladding panels are silkscreened with a repetitive image of a palm leaf photograph by Karl Blossfeldt.

The outer layer of Laban's double skin consists mainly of triple-layered polycarbonate panels measuring 40 millimeters (1-⅝ inch) in thickness and 500 millimeters (20 inches) in width. The panels, supplied by German manufacturer Rodeca, lock together at vertical joints in a tongue-and-groove configuration; there are no horizontal joints, as the panels extend the full height of the wall. Each panel is anchored for lateral support to a sub-frame consisting of galvanized steel-tube posts and anodized aluminum rails. The polycarbonate is technically transparent (although the multiple layering of each panel's interior cellular structure renders it translucent) except in certain areas where an integral custom-color tint is incorporated into the innermost surface of the panel. Gradated shades of magenta, turquoise, and green are used to mark the façade with the locations of dance studios within the building. The effect is subtle, as the blocks of color appear muted through the polycarbonate, with blurred edges reminiscent of

watercolor painting. At first suggesting a random arrangement, these colors are in fact each associated with particular studios, and the hues are repeated on selected corresponding interior walls and fixtures, becoming an orientation or way-finding device inside the building. Depending upon the exterior lighting conditions, the polycarbonate panels range in appearance from semi-translucent, allowing the viewer to see through to the inner wall, to translucent, indicating only a vague sense of depth behind the outer skin, to fully opaque, reflecting sunlight and appearing solid. In areas where direct views to the exterior are desired, the polycarbonate skin is intermittently replaced by large panes of reflective insulating glass, which are transparent from the inside by day but appear mirror-like from the exterior, presenting an echo of the mirrored walls found inside each dance studio. At night, with internal lighting, the reflective glass becomes transparent when viewed from outside, providing the only unimpeded glimpses of interior space.

The inner wall of the double-skin system is separated from the outer one by a space of 80 centimeters (2.6 feet), containing a steel-grate catwalk at each level for maintenance access. This cavity is vented along its base and above the roof to allow circulation of air via the stack effect. Where natural lighting is desired, the inner skin consists of double-pane insulating glass, typically spanning from floor to ceiling and supported by extruded aluminum

3.16   Dance studio interior. Laban Dance Center, Deptford, London, UK. Herzog & de Meuron, 2003.

3.17 Laban Dance Center, Deptford, London, UK. Herzog & de Meuron, 2003.

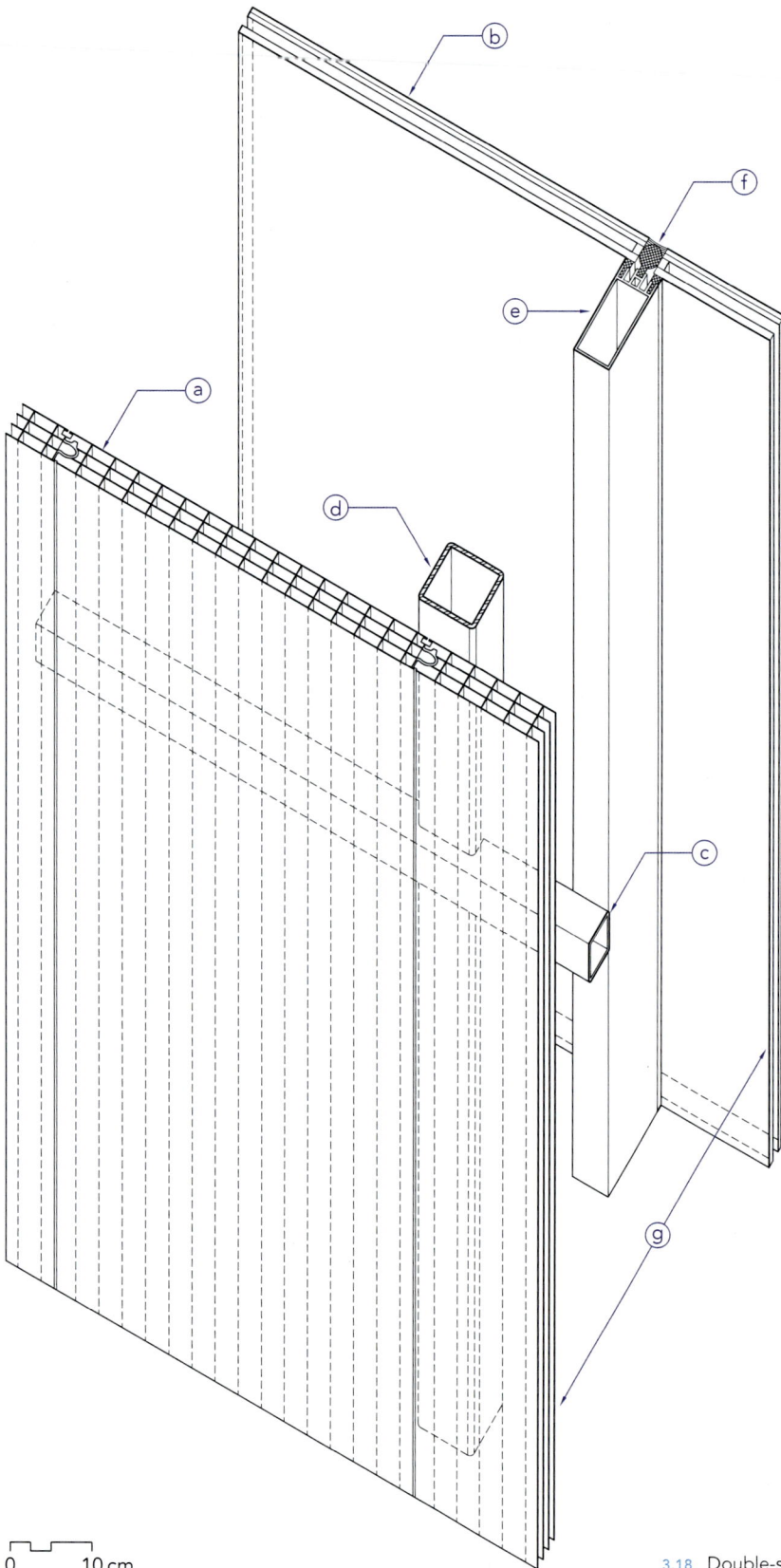

(a) Outer skin: translucent polycarbonate panel, 40 × 500 mm

(b) Inner skin: translucent laminated insulating glass unit, 38 mm thick

(c) Aluminum tube, 50 × 50 x 4 mm

(d) Galvanized steel tube, 80 × 80 × 5 mm

(e) Extruded aluminum mullion, 50 × 120 mm

(f) Silicone sealant

(g) Air cavity (depth not to scale)

0   10 cm

3.18   Double-skin wall assembly. Laban Dance Center.

3.19 Dance studio interior. Laban Dance Center, Deptford, London, UK. Herzog & de Meuron, 2003.

3.20 Laban Dance Center, Deptford, London, UK. Herzog & de Meuron, 2003.

mullions. A translucent matte film is applied to the glass to help further diffuse the sunlight that is transmitted through the outer polycarbonate panels. Within the dance studios, this wall of glass creates a serene environment, not only providing light but ensuring that outside distractions are minimized. Operable windows are incorporated into the glass wall to allow natural ventilation between the interior space and the air cavity. At least one transparent glass window (reflective from the outside) is provided for views from each studio.

In this double-skin system, the outer layer provides protection from glare as well as increased acoustic and thermal insulation, which, in conjunction with the air cavity and inner wall, lessens the demand for energy to heat and cool the building. The layering of polycarbonate paneling and translucent glass creates a luminous wall that ensures an abundance of natural light (and therefore reduced need for electrical lighting) and, more specifically, of appropriately diffuse lighting within each dance studio. At the same time, this assemblage of materials performs in other ways, which could be called its phenomenological functions. The

67

movement of clouds and sun across the sky are translated to the interior as shifting tones of shadow and light, infused by hints of color, linking the inhabitants' experience to the outside world without revealing it visually. The movements of dancers within the space are not put on display to the exterior (as transparent glass would do), but are registered as shadows against the translucent screen of the façade. These simple materials produce an ethereal beauty that is complicated and rooted in a specific place and time. In recognition of the Laban Center's complex layering of transparency, translucency, and reflection, and the projection of movement upon its façades, Giles Reid writes that the Laban skin creates a two-way mirror effect, making the building "a machine for observing movement, for heightening one's awareness of being watched, for becoming a performer within a social space."[14]

In the designs of the Steiff Factory and the Laban Dance Center, the combination of simple, rational form with an unusual material palette and the technologies of the double-skin façade underscore the primary importance of light and atmosphere to the interior life of each building and its program. Though initially conceived to provide straightforward solutions for their respective programmatic requirements, each building nevertheless produces subtle and fascinating side-effects of layering and depth, oscillating between concealment and revelation.

4
Compound Lens

STEUBEN GLASS BUILDING, NEW YORK CITY, USA, 1937

MAISON HERMÈS, TOKYO, JAPAN, 2001

In awarding the 1998 Pritzker Architecture Prize to Renzo Piano, the jury's citation praised the architect's prolific portfolio of buildings as "staggering in scope and comprehensive in the diversity of scale, material and form," adding that Piano "has remained true to the concept that the architect must maintain command over the building process from design to built work."[1] Although at the time Piano had not yet completed his design for Maison Hermès in Tokyo, the jury's comments seem prescient. Over more than three decades, Piano and his associates have developed a reputation for working closely with material manufacturers and fabricators to explore and execute new formulations and leading-edge applications for building materials; Maison Hermès is clearly a beneficiary of this process. Piano's office collaborated with the Italian glass manufacturer Vetroarredo to design the custom glass blocks that comprise the building's façade, as well as with Swiss fabricator Schmidlin to develop the steel framing system that supports the glass blocks. The building's envelope consists of 13,000 individual glass blocks, covering a total surface area of more than 2,600 square meters (28,000 square feet) and imparting the building's defining character: a unique form of translucency derived from light refraction through thousands of modular lens-like blocks.

The Renzo Piano Building Workshop (RPBW), as the firm is officially titled, is best known for its designs for cultural and high-rise projects around the world. These include Centre Georges Pompidou (1977) in Paris, the Menil Collection (1986) in Houston, the Potsdamer Platz Reconstruction (2000) in Berlin, Parco Della Musica (2002) in Rome, and the New York Times Tower (2007) in New York City.[2] Maison Hermès is the firm's third project in Japan, following the Kansai International Aiport (1994) and the Ushibuka Bridge (1996), but its first in Tokyo. Hermès, the fashion and luxury-goods manufacturer based in Paris, commissioned RPBW in 1998 to design its new flagship store in Tokyo's upscale Ginza district. The building occupies a corner site at the intersection of Harumi Dori, one of the district's busiest shopping avenues, and Sony Dori, a smaller side street. The site's slender proportions—just 11 meters wide by 45 meters long (36 by 148 feet)—are

4.1   Maison Hermès, Tokyo, Japan. Renzo Piano Building Workshop, 2001.

70

4.2   New York Times Tower, New York City, USA. Renzo Piano Building Workshop, 2007.

4.3   Harumi Dori, Ginza district, Tokyo, Japan. Maison Hermès at center.

typical for Ginza. The building was completed in 2001 and contains 6,000 square meters (65,000 square feet) of space on 12 floors above ground and three below, with a total height of 45 meters (148 feet).[3] The program includes retail space, a design atelier, office space, a small cinema and museum, a rooftop garden, and a subterranean link to the Tokyo subway. In plan, a spine of circulation and service spaces runs along the party wall abutting the neighboring building, with open retail and office spaces occupying the remainder of the narrow floor plates along the perimeter glass walls.

Facing the challenge of designing a building façade that could attract desired public attention amidst a commercial district typified by colorful signage, massive billboards, and dazzling light shows, RPBW chose a strategy of minimalist restraint—to stand out by standing back.[4] The main façades of Maison Hermès consist of a single repeating module: the 45-centimeter square (17-¾ inch) by 12-centimeter thick (4-¾ inch) translucent glass block. The only two exceptions to this rule are at the corners, where smaller, curved glass blocks (one-quarter the size of the typical block) are used to wrap the rounded edges of the building, and at the street-level retail entrance, where some large transparent-glass display windows are provided. Additionally, at eye level for pedestrians on the sidewalk, small transparent display cases (similarly dimensioned as the glass blocks) are occasionally interspersed among the typical blocks.

The regularity of the façade's simple grid belies its complex tectonics. The custom-fabricated glass blocks of Maison Hermès are much larger than standard glass blocks, which are typically just 30 centimeters square (12 inches) or smaller. Each glass block is individually supported by a concealed network of thin steel channels embedded within the joints between blocks, a system designed to accommodate the anticipated seismic movement stipulated by Tokyo's stringent codes (and no mortar is used, as would be typical in standard glass-block construction). If a seismic event induces building movement at Maison Hermès, each glass block can absorb its share, up to 4 millimeters per side. The cross-sectional profile of the glass block is designed to provide a cavity along each edge to house these steel channels (which measure approximately 53 by 80 millimeters, or 2 by 3 inches). The grid of

4.4 Maison Hermès, Tokyo, Japan. Renzo Piano Building Workshop, 2001.

45 cm

(d)

(e)

(e)

(c)

(a)

(b)

(a) Custom hollow glass block,
45 × 45 × 12 cm

(b) Silicone sealant

(c) Steel channel, 53 × 80 mm

(d) Steel channel, 18 × 80 mm

(e) EPDM-rubber sleeve

0   10 cm

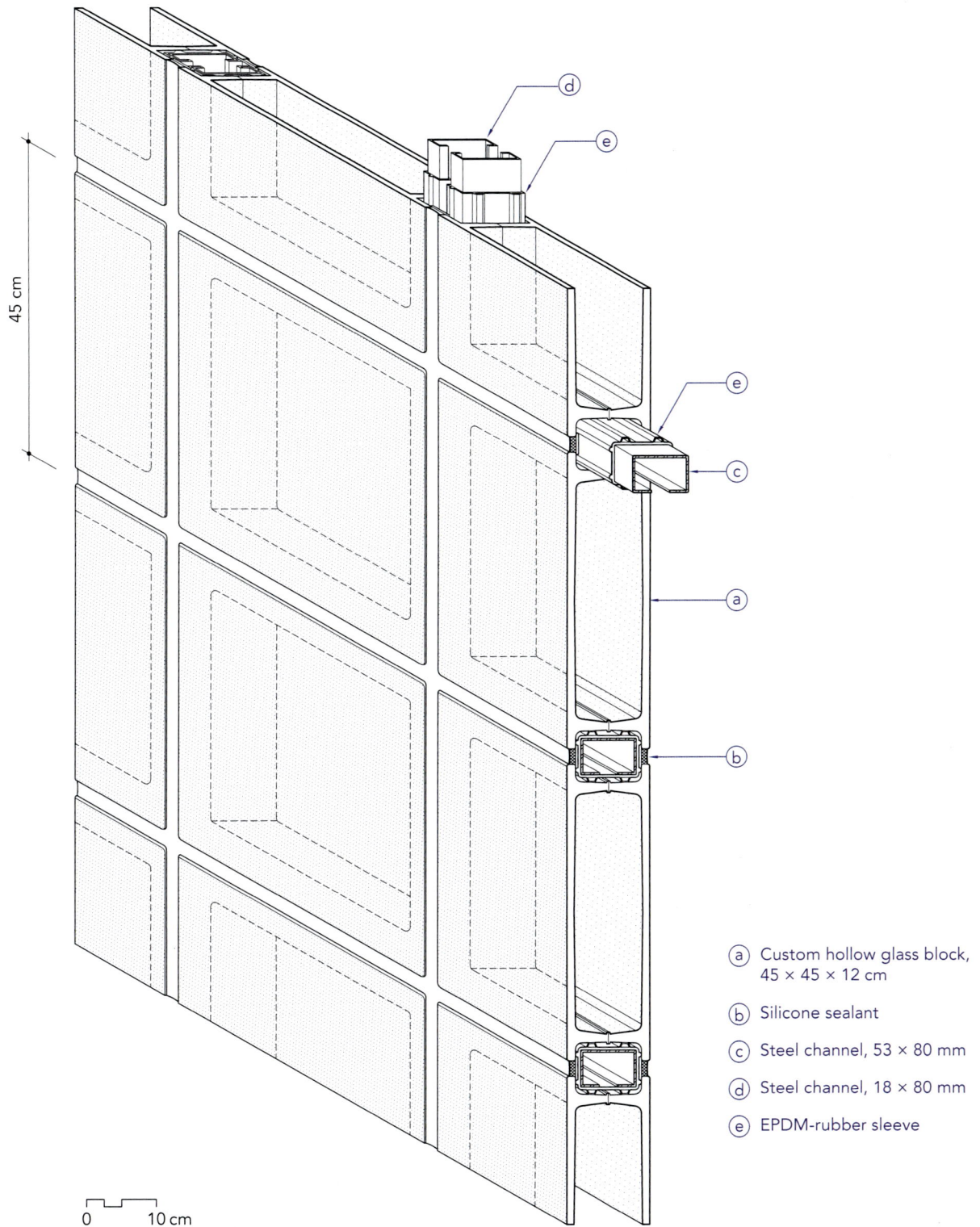

4.5 Glass-block curtain wall. Maison Hermès.

4.6 Maison Hermès, Tokyo, Japan. Renzo Piano Building Workshop, 2001.

4.7   Maison Hermès, Tokyo, Japan. Renzo Piano Building Workshop, 2001.

4.8   Interior, Maison Hermès, Tokyo, Japan. Renzo Piano Building Workshop, 2001.

4.9 Maison Hermès, Tokyo, Japan. Renzo Piano Building Workshop, 2001.

4.10　Maison Hermès, Tokyo, Japan. Renzo Piano Building Workshop, 2001.

channels is suspended at each level from a wide-flange steel beam running along the edge of each cantilevered floor slab. An EPDM (ethylene propylene diene monomer) rubber sleeve surrounds each channel, preventing direct glass-to-metal contact and providing accommodation for expansion and movement of both glass and metal. The resultant joint between each glass block is approximately 20 millimeters wide (¾ inch) and filled with elastic silicone sealant.

Glass blocks are typically hollow, fabricated from two dish-like pieces which are fused together under intense heat. The manufacturing process begins with the heating of base ingredients, including recycled glass, sand, soda ash, and lime, until molten. A precise amount of this mixture is then pressed into a mold and cooled with a blast of air to solidify it, creating a glass piece with one large surface and four raised edges—this is one half of the block. The surfaces of the mold may contain a textured pattern which is then impressed onto the surface of the glass. To fabricate the finished block, two halves are reheated until their edges begin to soften and are then pressed together and cooled again, thus

fusing the two halves into one while forming a sealed interior air chamber that gives the glass block its thermal and acoustical insulating qualities. The two halves of a glass block are essentially plano-concave lenses, and therefore light rays passing through them converge and diverge at various angles, producing a semi-transparent, semi-translucent effect depending upon surface textures and angles of view. The exterior surfaces of the custom blocks at Maison Hermès have a smooth hand-polished finish, while the interior surfaces have a variegated texture. Each glass block within the building skin therefore acts like a lens through which the passage of light is manipulated in both directions (sunlight from the exterior by day, interior illumination at night), projecting an abstraction of shapes, shadows, color, and movement to the other side. This is particularly effective along the building's long elevation on Sony Dori, where open staircases running adjacent and parallel to the glass wall on the interior convey a sense of diagonal movement to passersby outside. The overall effect of translucency thus achieved gives the Maison Hermès a sense of intrigue, providing hints of activity beyond the

4.11 Maison Hermès, Tokyo, Japan. Renzo Piano Building Workshop, 2001.

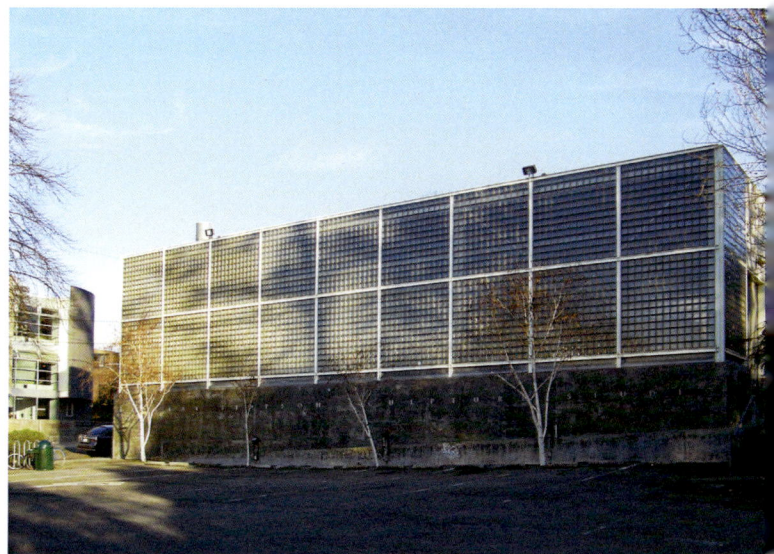

4.13 Barclay Simpson Sculpture Studio, Oakland, California, USA. Jim Jennings, 1992.

4.12 Academy of Arts, Maastricht, the Netherlands. Wiel Arets Architects, 1993.

glass walls and inviting further exploration and contemplation. Piano has described the building's envelope as both a magic lantern and a glass veil that creates "a continuous luminous screen between the serenity of the inner spaces and the buzz of the city" and "leaves more to the imagination than can actually be seen."[5]

Similar effects can be found in an earlier building in the Netherlands by Wiel Arets Architects: the Maastricht Academy of Arts, completed in 1993. The project includes the refurbishment of an existing structure and the addition of two buildings housing workshop spaces for architecture, painting, sculpture, and fashion, in addition to an auditorium and library. Diffuse natural light is brought to the interior through a building skin of continuous translucent glass blocks, interrupted only by metal frames that express the structural grid behind it and, within each structural module, a band of transparent glass with an operable window to provide natural ventilation and views. Unlike Maison Hermès, the Maastricht Academy uses standard glass blocks, assembled in the traditional manner, but still manages to achieve a singular expression through rigorous attention to detailing and an articulation of

4.14 Christian Dior
Building, Tokyo,
Japan. SANAA, 2003.

4.15 Louis Vuitton Store, Tokyo, Japan. Jun Aoki, 2004.

scale translated from the glass-block grid to the building-structure grid. On a smaller scale, the same issues are explored in architect Jim Jennings's design for the Barclay Simpson Sculpture Studio (1992) in Oakland, California, where a concrete base supports structural steel members that frame walls of translucent glass blocks, admitting an even, diffuse light to the artists' spaces within.

One phenomenon that will surely define a trend of early twenty-first-century architecture was the tendency for European fashion and luxury-goods houses to hire well-known architects to design iconic flagship stores in urban centers, particularly in Japan.[6] In the interest of placing Maison Hermès within a contemporary context, it is also useful to look at two other such buildings within the city of Tokyo which were completed shortly after Hermès and which use very different techniques to arrive at similar goals. These buildings—the Christian Dior Building (2003) in the Omotesando district, designed by Kazuyo Sejima and Ryue Nishizawa of SANAA, and the Louis Vuitton Store (2004) in Rappongi Hills, designed by Jun Aoki—deploy glass (though not glass blocks) and other materials in innovative ways to develop unique visual and lighting effects that are integral to their architecture rather than being applied as signage. Like Maison Hermès, the enclosure system of each building relies on a limited material palette and minimalist detailing.

The envelope of SANAA's Dior Building consists of two layers: an outer wall of transparent flat glass panels spanning from floor to floor, without mullions, and an inner wall made of custom translucent acrylic panels vacuum-formed into subtly flowing shapes intended to recall fabric pleats or the movement of silk drapes. The acrylic panels are also printed with a pattern of white lines that work in combination with each piece's irregular curvature to produce a moiré pattern that changes with lighting conditions and angles of view. The resultant screening effect allows views through the panels from inside to out, but creates a mysterious cloud- or mist-like appearance from the outside.

The unusual effect created by Jun Aoki's design for his Louis Vuitton building skin is best characterized as kaleidoscopic translucency. The main portion of the street-front façade consists of an utterly unique and complex system of components, including 28,000 cylindrical glass tubes arranged like cells in a honeycomb and inserted in holes cut into reflective stainless steel sheets, all of which are encased between two sheets of transparent flat glass. Each glass tube measures 10 centimeters (4 inches) in diameter and 30 centimeters (12 inches) in length. Variations in tube placement spell out the name Louis Vuitton across the length of the main façade. When viewing this system directly from a frontal perspective, one may experience a brief glimpse of the interior through a few tubes while simultaneously perceiving reflections from obliquely viewed tubes and the stainless steel sheet surrounding them. From a distance, the tube system reads like a solid yet porous block of ice, or perhaps a mass of pixels on a screen, hovering over a narrow band of transparent display windows. It is through the multiplication of transparent and reflective surfaces that a dramatic new form of phenomenal translucency is achieved, creating a building façade that engages us with both literal and illusory depth.

* * *

Although the translucent building skin of Maison Hermès has been frequently compared to Pierre Chareau's Maison de Verre of 1932, there is another lesser-known precedent—a commercial building in New York City—which is worthy of further examination.[7] The Steuben Glass Building was constructed in 1937 (nearly contemporary with Maison de Verre) at the southwest corner of Fifth Avenue and 56th Street in Manhattan. Designed by architect brothers William and Geoffrey Platt,[8] the Steuben Glass Building shares many attributes with Maison Hermès: both structures house office and commercial display spaces, both occupy prominent corner sites in bustling urban settings, and both feature translucent glass-block façades built on an unprecedented scale.

Steuben was the art-glass division of Corning Glass Works and was known for its production of fine glassware and decorative objects. The company's factory was located up-state in Corning, New York, and it had retail outlets in major cities around the country. The new Steuben Glass Building (sometimes referred to as the Corning Building) had a distinctive late Art-Deco character and stood six stories tall, containing show rooms for Steuben

Fifth Avenue façade, Steuben Glass Building, New York City, USA.
William and Geoffrey Platt, 1937.

4.17 Steuben Glass Building, New York City, USA. William and Geoffrey Platt, 1937.

30 cm

30 cm

(a)

(b)

(a) Translucent hollow glass
block, 30 × 30 × 10 cm

(b) Mortar

(c) Limestone cladding on brick

(d) Formed metal frame

0   10 cm

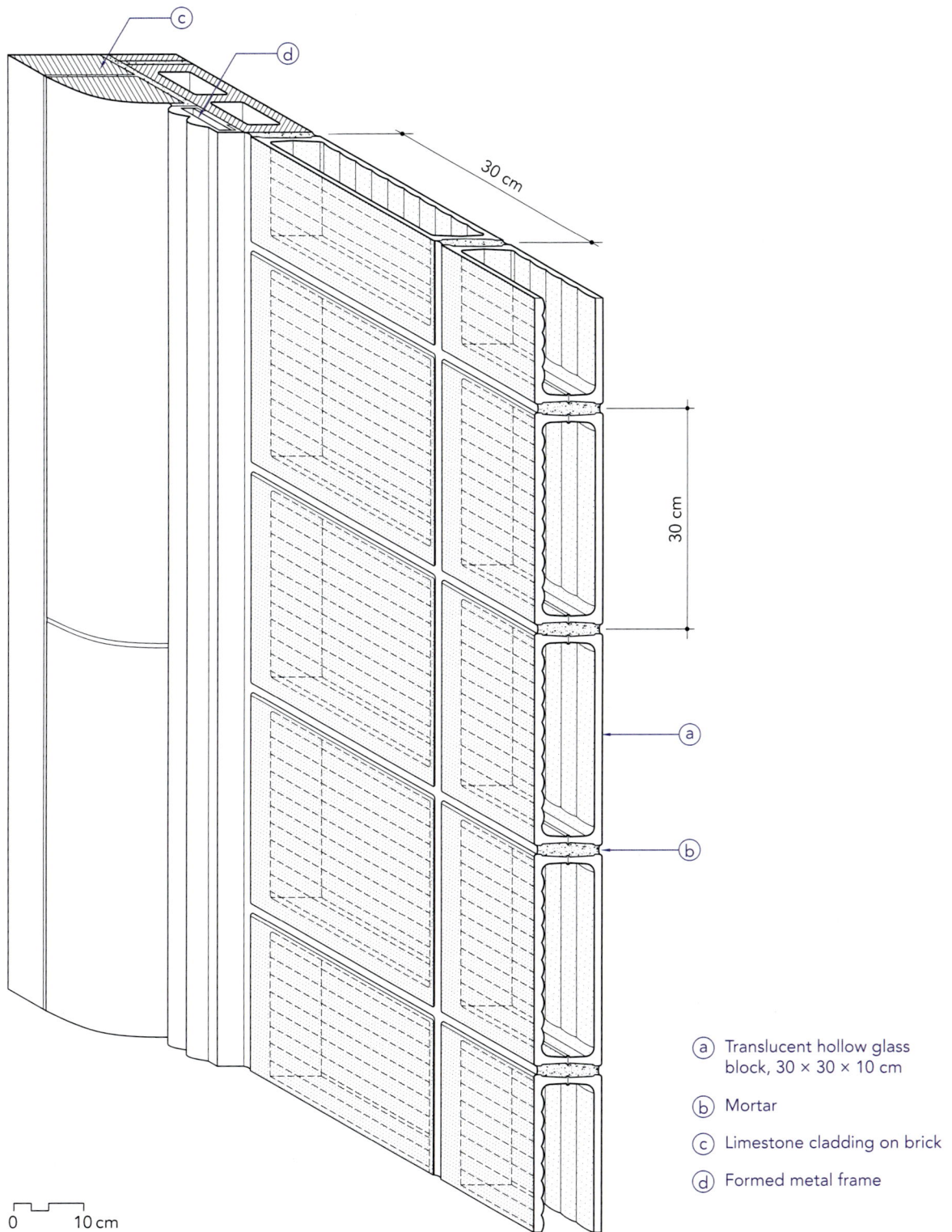

4.18 Glass-block wall assembly. Steuben Glass Building.

4.19　Office interior, Steuben Glass Building, New York City, USA. William and Geoffrey Platt, 1937.

products on the ground level with office spaces above. The site was 8.2 meters wide by 30.5 meters long (27 by 100 feet), which happens to be quite similar in proportion to the site of Maison Hermès. The building contained approximately 1,500 square meters (16,000 square feet) of interior space.

In a 1937 review of the newly completed Steuben Glass Building, the critic Lewis Mumford wrote: "The general elements in this structure are simple and excellent. In a frame of broad Indiana limestone slabs, two great vitreous masses are placed."[9] These vitreous masses were in fact four-story curtain walls—a narrow one facing Fifth Avenue and a broad one facing 56th Street—comprising 3,800 translucent glass blocks, used to the exclusion of transparent windows (with the exception of a few display windows at street level). Manufactured by Corning, the glass blocks each measured 30 centimeters square (12 inches) and 10 centimeters thick (4 inches). The interior face of each half of a block had a fluted surface texture, with the flutes running vertically on one face and horizontally on the other, producing a crisscrossing pattern that obscured vision and prevented glare but allowed transmission of soft, diffuse light. The fluted surfaces each faced inward toward the air cavity within the block, leaving the outer

surfaces smooth and therefore easily washed (a feature commented upon in several press reports at the time). The opaque portions of the façade were restricted to a perimeter frame around the glass block areas and were constructed of brick masonry clad with limestone panels. Brick masonry was also used in front of structural columns and in horizontal bands, 90 centimeters deep (3 feet), spanning along the edge of each floor slab to provide required fire-separation between levels. These brick bands were concealed behind glass blocks but were also expressed via a grid of nickel-silver framework on each façade. Construction was streamlined, as the glass units were installed by the same masons who built the brick walls, utilizing a similar Portland-cement mortar to form the 6-millimeter (¼-inch) joints.

In addition to providing display space for Steuben's products, the building presented a unique opportunity for the architecture to become a literal representation of its owner's identity through the extensive use of construction materials manufactured by Corning. Glass blocks were used not only on the façade but also as interior partitions, allowing "borrowed" light to reach deeper into the building and not only into perimeter offices. Glass fiber was used for thermal and acoustic insulation as well as to wrap

4.20 56th Street façade, Steuben Glass Building, New York City, USA. William and Geoffrey Platt, 1937.

4.21 Corning Glass Tower, New York City, USA. Wallace K. Harrison, 1959. Note Steuben Glass Building at right.

4.22 Corning Glass Tower. Note renovated façade of Harry Winston Building (formerly Steuben Glass Building) at right.

plumbing pipes. Glass tiles, balusters, and mirrors were used as decorative interior finishes. The greatest impact in the use of glass, however, was achieved in the building enclosure, which enabled light-filled interior spaces throughout and created a dazzling, urban-scaled exterior display of the realized potential of modern glass architecture. Remarking on the decision to use glass block as the primary building envelope material, Geoffrey Platt said in 1936, "It has freed the architect from the feeling of a wall as a solid, opaque mass which encloses the building and through which he has to punch holes for windows. He is now able to think of his exterior wall as a screen, transmitting light."[10]

The Steuben Building façade, as designed by William and Geoffrey Platt, unfortunately no longer exists. In fact, it survived just 23 years. In 1959, Corning Glass Works built a new 28-story headquarters directly across Fifth Avenue from the Steuben Building. Designed by Wallace K. Harrison, who had previously collaborated on the design of the United Nations complex in New York, the new tower, with its own notable glass curtain wall, consolidated all of Corning's Manhattan operations, including the Steuben division.[11] The Steuben Building was then sold to the

jeweler Harry Winston, whose company continues to occupy the building today. Winston, working with architect Jacques Régnault, demolished all of the glass blocks and limestone panels in order to rebuild the façade in a bland eighteenth-century French style with traditional punched windows and travertine cladding and cornices.[12] In a sadly ironic twist, the jeweler replaced an unusual gem-like building façade of historic significance with a generic one that merely imitates a historical style. Ada Louise Huxtable was particularly critical of Winston's renovation, calling it "architecture as play acting" and writing that accepting an imitation as the real thing "wouldn't work with Mr. Winston's jewels, and it doesn't really work with architecture either."[13]

Although a 1936 New York Times report on the soon-to-be constructed Steuben Glass Building called it "New York's first glass house,"[14] there was in fact a recently renovated house about ten blocks away that was in many ways a precursor of the Steuben Building. This building was an actual residence: the Lescaze Townhouse and Office (1934) at 211 East 48th Street. Designed by William Lescaze to house his family and his architectural office, the four-level structure featured a façade largely built of 125

4.23 Harry Winston Building (formerly Steuben Glass Building).

been called by historians "New York's preeminent example of the International Style."[15] In recognition of its historical significance, the Lescaze Townhouse was listed in 1980 on the US National Register of Historic Places.

Another predecessor to the Steuben Glass Building can be found in Frank Lloyd Wright's unbuilt 1897 design for the Luxfer Prism Skyscraper. Wright prepared a perspective drawing of his theoretical project to accompany the announcement of a design competition sponsored by the Luxfer Prism Company, a manufacturer of glass prism tiles and panels.[16] Luxfer's tiles were not glass blocks but performed in a similar way. The glass tiles were flat on one face and ridged on the other, and could thereby act as prisms within a façade to redirect natural light deeper into a space (a typical application was in high transom windows within storefronts). Wright envisioned a ten-story tower with a building skin composed primarily of translucent Luxfer prism tiles, spanning from floor to ceiling on each level, with the only opaque elements being narrow bands of stone or terracotta cladding covering the building's structural frame. Luxfer prism tiles were used in another influential project that was constructed 17 years later: Bruno Taut's expressionist Glass Pavilion, built for the 1914 Werkbund Exhibition in Cologne, Germany. In fact, the German branch of the Luxfer Prism Company sponsored the construction of the pavilion to serve as a demonstration of its products. Circular in plan and containing two floor levels, the building had a reinforced concrete base supporting glass walls topped by a pointed glass-tiled dome. Glass was also used in interior partitions, floors, and even a staircase. Detlef Mertins describes the structure as "a total work of art that integrated glass construction, art, and mosaics, and that induced an altered state of consciousness."[17] While diminutive in scale, the Glass Pavilion has become an important symbol of early twentieth-century modern architecture. Though they were exhibited to great fanfare at Cologne, glass prism tiles were essentially made obsolete less than two decades later, with the development in the 1930s of hollow glass blocks by Corning and others. Glass blocks could achieve the prismatic effects of Luxfer tiles and had the added benefits of a sealed air cavity for insulation, both thermal and acoustic, and the capability of being stacked in load-bearing walls.[18]

millimeters square (5 inches) translucent glass blocks on the upper two floors. It represented an assertively new presence in Manhattan, contrasting starkly with its traditional nineteenth-century brownstone neighbors. The glass blocks were produced by the Macbeth-Evans Glass Company of Pennsylvania, which later merged with Corning Glass Works. Lescaze, who is best known for the Philadelphia Savings Fund Society tower (1932) in Philadelphia, was a partner with George Howe in the firm of Howe & Lescaze. The pair also produced a series of unbuilt designs for a new Museum of Modern Art (see Chapter 2). With its crisp form, white stucco, and extensive glazing, the Lescaze Townhouse has

4.24 Lescaze Townhouse and Office, New York City, USA. William Lescaze, 1934.

In the field of optics, a compound lens is broadly defined as an array of devices which act together to transmit and refract light in a specified pattern. A glass-block façade performs in a similar way, creating unique effects through the refraction of light—effects which can have a profound impact on the experiential qualities of a building's interior spaces by day and the surrounding exterior spaces at night. The Steuben Glass Building and Maison Hermès each pushed the use of glass blocks to unprecedented heights, creating curtain walls of modular glass units that aggregate to form massive urban-scaled compound lenses.

4.25 Glass Pavilion, Werkbund Exhibition, Cologne, Germany. Bruno Taut, 1914.

5
Geology

Reflecting upon his design for the Beinecke Rare Book Library at Yale University 25 years after its completion, Gordon Bunshaft said, "I think it is one of the half-dozen best buildings I've ever done in my life. It's the only building I've been involved in that has an emotional impact."[1] This singling-out of the Beinecke Library is remarkable in several aspects. First, Bunshaft had a long career as a designer at one of the twentieth century's most prolific architecture firms—Skidmore, Owings & Merrill (SOM). As a partner at SOM's New York City office, Bunshaft was responsible for designing many of the institutional and corporate buildings that came to define an important segment of postwar modernism in America.[2] Among his influential works were Lever House (1952), Manufacturers Hanover Trust Company (1954), Pepsi-Cola Headquarters (1960), Chase Manhattan Bank (1961) in New York City, and the Hirshhorn Museum (1974) in Washington, DC.[3] The Beinecke Library, at 8,200 square meters (88,000 square feet) in area, is a relatively small project located outside of New York City, where many of Bunshaft's larger, well-known projects were built. While much of his work was engaged primarily with the dual aesthetic obsessions of midcentury modernism—pure, simple form and transparency—the Beinecke Library embraces only the former while renouncing the latter to focus instead on the experiential qualities of translucency, the intangible source of the "emotional impact" that, for Bunshaft, characterized the Beinecke Library.[4]

Another striking fact about this building is that it produces its translucent effects not through the typical material of glass but instead through a unique formulation of stone cladding. Whereas stone and glass have traditionally been conceived as architectural opposites (heavy vs. light, thick vs. thin, opaque vs. transparent), in the Beinecke Library—as well as the other project featured in this chapter, Kengo Kuma's LVMH Osaka building—the accepted characterization of stone and its properties is subverted: stone instead becomes analogous to glass, a thin membrane that transmits light.

5.1 Beinecke Rare Book Library, New Haven, Connecticut, USA. Gordon Bunshaft, SOM, 1963.

5.2 Manufacturers Hanover Trust Company, New York City, USA. Gordon Bunshaft, SOM, 1954.

5.3 Beinecke Rare Book Library, New Haven, Connecticut, USA. Gordon Bunshaft, SOM, 1963.

5.4 Interior view from mezzanine level.
Beinecke Rare Book Library, New
Haven, Connecticut, USA. Gordon
Bunshaft, SOM, 1963.

265 cm

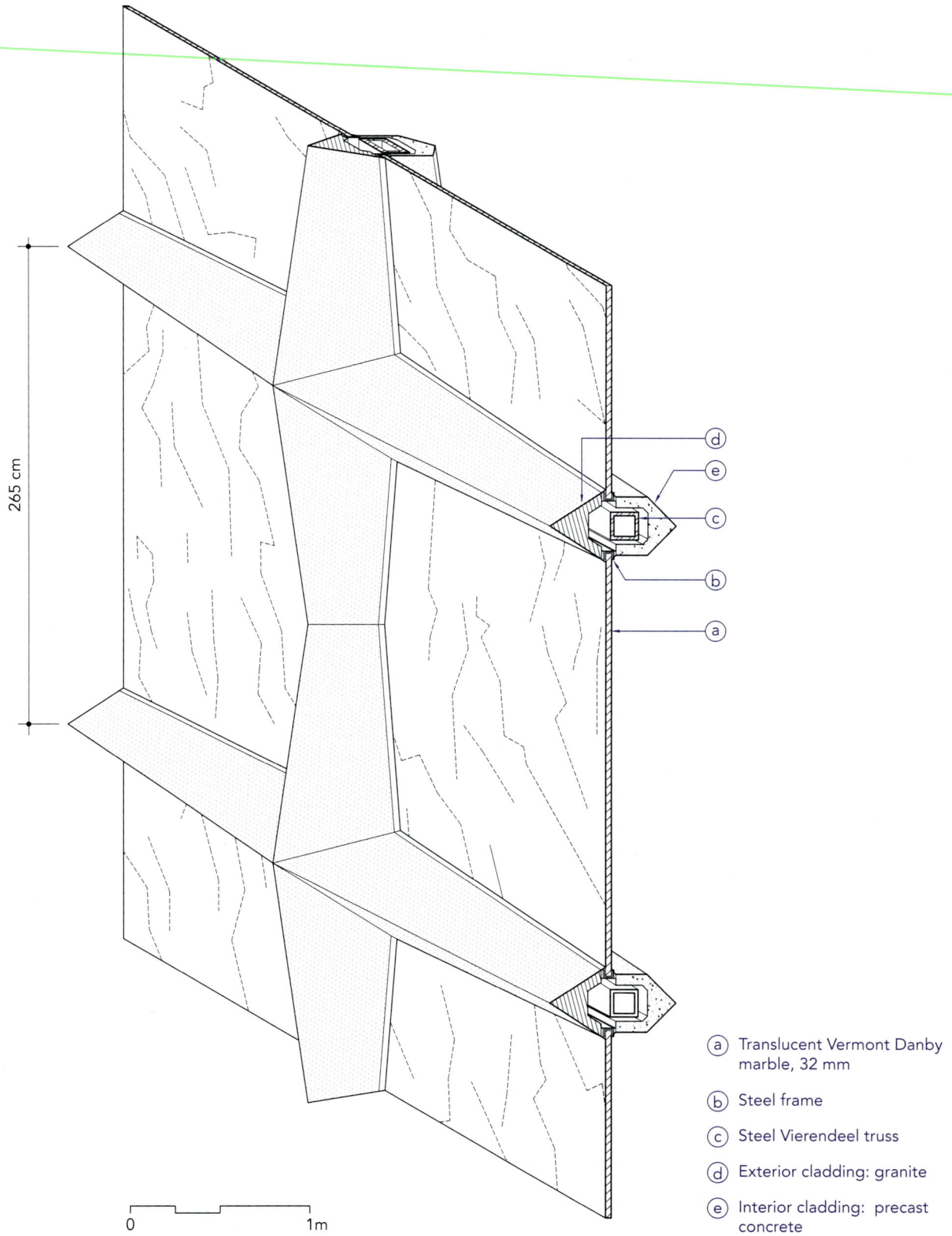

(a) Translucent Vermont Danby marble, 32 mm

(b) Steel frame

(c) Steel Vierendeel truss

(d) Exterior cladding: granite

(e) Interior cladding: precast concrete

0   1m

5.6   Wall assembly. Beinecke Rare Book Library.

5.7 Construction of the Beinecke Rare Book Library.

The library's construction and endowment were funded by Edwin, Frederick, and Walter Beinecke, three brothers and Yale University alumni who also donated their own significant collections of rare books and manuscripts to Yale. Originally other notable architects were also under consideration (including Eero Saarinen and Edward Durell Stone) but Bunshaft won the commission to design the library, which was to include space for 800,000 books, as well as a reading room, an exhibition hall, and staff offices. Construction began in 1960, and the building was completed in 1963. In his 1970 book *Great Libraries*, Anthony Hobson calls the Beinecke Library "the most imaginative construction of its kind for at least two centuries."[5]

Bunshaft's design divides the program into two distinct realms: first, an elevated translucent stone-clad rectangular volume that appears to float above a ground-floor entrance lobby and plaza, and second, a subterranean zone of two levels organized around a sunken courtyard. The below-ground levels house offices, reading spaces, and storage vaults for most of the book collection; the courtyard brings light to these lower floors and contains three abstract stone sculptures by Isamu Noguchi. The building's upper volume, measuring approximately 27 × 40 meters (88 × 131 feet) in plan and standing 18 meters tall (60 feet), is dedicated entirely to the storage, display, and celebration of books. Within this volume is a 13.7-meter high (45-foot) cathedral-like space with a mezzanine level, containing exhibition and lounge spaces, which surrounds a freestanding glass cube. This six-level glass-enclosed reliquary holds shelving for 160,000 books (the remainder is stored in below-ground vaults). The bronze-framed curtain wall of this inner book-stack tower was designed to maintain an optimum interior temperature of 21 degrees Celsius (70 Fahrenheit), with 50 percent relative humidity, to protect the books, while also placing their colorful spines on display through large panes of transparent glass. The building's exterior walls, which reveal little of the interior richness from the outside, are composed of a grid of translucent marble panels that bathe the interior spaces with diffuse natural light that ranges from subtle to dramatic in character.

Aside from the recessed gray-tinted glass walls of the ground-floor entry and lower court, the main exterior walls of the library are formed by steel Vierendeel trusses which support the roof and span the full length of each side of the building, resting only on four corner columns at the plaza level. The vertical and horizontal members of the trusses are clad with granite on the exterior and with precast concrete on the interior and are expressed as a series

5.9 Interior view of marble cladding at mezzanine. Beinecke Rare Book Library.

5.10 Preliminary model of the Beinecke Rare Book Library.

5.11 Beinecke Rare Book Library, New Haven, Connecticut, USA. Gordon Bunshaft, SOM, 1963.

of octagonal frames which support 250 panels of Vermont Danby marble. Each of these panels measures about 2.4 meters (8 feet) in width and height, with a thickness of just 32 millimeters (1.25 inches), and is individually anchored to the steel truss members by steel anchor brackets. It is the remarkable thinness of each marble slab that produces its translucency and allows diffuse daylight to be transmitted to the interior, while blocking the passage of ultraviolet light, which is potentially damaging to paper and fabrics. The thin marble slabs, however, also have inherently poor performance in terms of thermal insulation, therefore requiring the inner, secondary glass enclosure for the books. From the exterior, the white surface of the marble skin appears opaque in daylight, echoing the solidity of the neighboring stone buildings and plaza. The muteness of the exterior is starkly contrasted by the interior, where each marble slab, backlit by the sun, reveals its unique veining and coloration. In fact, when struck by direct sunlight, the marble is ablaze on the interior with unexpected variation in tone, from dark, earthy browns to amber and streaks of white. Though direct views to the exterior are prevented, through the shifting of light and shade the visitor develops an awareness of directionality, of the movement of the sun, and of the passage of clouds.

During the design process, several different materials had been considered for the Beinecke's translucent building skin. Bunshaft originally favored onyx, inspired by a building he had seen years before on a trip to Istanbul. During the design phase, the architects of SOM were in communication with onyx suppliers in Peru who initially claimed to be capable of providing adequate numbers of the large onyx sheets required for the project, but ultimately failed to produce panels of the desired quality. Alabaster and translucent glass were also briefly under consideration as alternatives before Danby marble from nearby Vermont was deemed most suitable for the project.

Stone used in building construction is typically categorized as igneous (such as granite and basalt), sedimentary (such as limestone and sandstone), or metamorphic (such as marble and slate). The distinctions among these types relate to how the stone was originally formed: whereas *igneous* rock was deposited and cooled from a molten state, and *sedimentary* rock formed from particles deposited in water, *metamorphic* rock results from

5.12 Beinecke Rare Book Library, New Haven, Connecticut, USA. Gordon Bunshaft, SOM, 1963.

further transformation through heat and pressure of material that was formerly either igneous or sedimentary. Metamorphic stones, like marble (which is a recrystallized form of limestone), often display significant veining of various colors resulting from the inclusion of additional minerals mixing with the original rock during the metamorphic process. To achieve translucency with marble, however, the problem of fabricating a slab thin enough to allow light transmission was a major obstacle. Advancements in fabrication tools and techniques developed in the twentieth century, including precise diamond-blade saws, provided the technology necessary to produce slabs of unprecedented slenderness (down to a fraction of an inch), thus making it possible to realize the translucent stone skins of buildings like the Beinecke Library and others that followed it.

The advent of thin veneer-stone cladding techniques brought with it some significant problems, however. Chief among these is the phenomenon of thermal hysteresis, or the permanent physical deformation of a material resulting from exposure to variations in temperature. Italian Carrara marble, for instance, is subject to hysteresis when used as thin cladding panels in climates with wide temperature fluctuations. When exposed to weather over time, Carrara panels can transform in shape from a flat plate to a concave or convex dish. As this thermal-cycling process continues, the panels will eventually lose flexural strength and may dislodge from their anchoring system—obviously a serious danger when used on the façades of tall buildings. Two famous stone façade failures brought the issue of thermal hysteresis to the forefront of building technology discourse. Edward Durell Stone's 83-story Amoco Building (now Aon Center) in Chicago, completed in 1973, was clad with 44,000 panels of Carrara marble that ranged in thickness from 3 to 4 centimeters (1-¼ to 1-⅜ inch). Within ten years, visible outward bowing of the stone panels was evident. Testing indicated that the marble had lost 40 percent of its flexural strength and was in danger of failing. By 1991, all of the marble panels were removed and replaced with 38-millimeter thick (1-½ inch) panels of North Carolina granite, a stone similar in appearance to marble but not subject to hysteresis. Alvar Aalto's Finlandia Hall, built in 1971 in Helsinki, suffered a similar fate. Within twenty years, panels of 3-centimeter thick (1-¼ inch) Carrara marble began to cup and fall from the façade and were eventually completely replaced in 1997 at a cost of 3 million Euros. Due to the importance of preserving Aalto's original design intent, however, the marble panels were not replaced with granite or another material immune to hysteresis—they were replaced with similar white marble that began showing similar signs of deformation within three years.[6] Fortunately for Bunshaft and Yale, the Vermont

5.13 Amoco Building (now Aon Center), Chicago, Illinois, USA. Edward Durell Stone, 1973.

Danby marble selected for the Beinecke Library is not susceptible to thermal hysteresis (a fact that was likely unknown at the time of construction) and remains stable when used as thin veneer cladding.

The result of Bunshaft's unusual design for the Beinecke building skin and its interaction with light is an interior experience characterized by a meditative, ethereal, and even spiritual effect that is not only appropriate for the reflective study of rare books

5.14 St. Pius Church, Meggen, Switzerland. Franz Fueg, 1966.

but also provides the emotional component of experience described by Bunshaft.[7] It is not surprising, therefore, that similar material strategies can be found in the subsequent design of religious spaces by a range of architects. The St. Pius Church at Meggen, Switzerland, for example, was designed by Franz Fueg and completed in 1966, three years after the Beinecke Library. A simple rectangular volume encloses a main worship space and altar, wrapped on four sides by walls of translucent stone. The building envelope incorporates marble panels of a similar thinness

(28 millimeters) to the Beinecke Library and likewise offsets its amber-toned luminous stone with a clearly expressed structural system, in this case steel columns which are integral with the walls and support the exposed steel trusswork of the roof.[8] St. Pius is a modern, abstract reinterpretation of the tradition of stained glass in religious architecture, utilizing the concept of embodied light as the representation of divine presence in the church. A similar approach is taken by architects von Gurkan, Marg & Partners in their design of the Christ Pavilion, completed in 2000 in

5.15 Cathedral of Our Lady of the Angels, Los Angeles, California, USA. Rafael Moneo, 2002.

Volkenroda, Germany. In this case, however, much thinner marble panels—just 10 millimeters thick—are laminated to 11-millimeter thick glass panes for structural support, producing similarly diffuse light, an inward focus, and a contemplative atmosphere.

Rather than creating an all-encompassing translucent skin like the previous examples, at the Cathedral of Our Lady of the Angels (2002) in Los Angeles, architect Rafael Moneo uses thin stone panels to strategically infuse natural light into certain areas within the cathedral, notably including a large three-dimensional concrete cross located above the altar and surrounded by translucent windows and skylights. The use of these translucent panels, consisting of 15-millimeter thick (⅝ inch) alabaster slabs laminated to glass, recurs in a series of secondary chapels, thus developing a narrative of light that unifies one's experience of the building's interior and contributes to its pervasively meditative character. It also suggests an abstracted reinterpretation of the tradition of stained glass in cathedral architecture. Moneo describes the interior as "enveloped in a luminous atmosphere that transports us far from the outside world, into a realm that we associate with the sacred . . . giving a spatial experience close to that encountered in certain Byzantine churches."[9]

* * *

Having transitioned from the secular to the sacred, we now return again to the secular realm. As opposed to the institutional and religious spaces discussed above, architect Kengo Kuma's design of the LVMH Building (2004) in Osaka, Japan, exploits the unique aesthetic effects of translucent stone within the everyday realm of commercial urban life. Sited not within an idyllic plaza on a university campus but rather in a dense, hyperactive metropolitan center, Kuma's design uses translucency as a means to create architectural identity. The nine-story building contains 8,300 square meters (90,000 square feet) of interior space, including boutique shops on the lower three levels and six floors of offices above. Built for the Louis Vuitton Moet Hennessy Group, the building benefits from a prominent location along one of Osaka's most important shopping boulevards but must compete for attention amongst the cacophony of signage and lighting that

5.16 LVMH Building, Osaka, Japan. Kengo Kuma and Associates, 2004.

characterizes the district. Kuma's strategy is to stand out through restraint.

By day, the building appears as an opaque, finely proportioned, minimalist stone cube perched atop a mute base of metal panels and glass. At night the building's translucency is revealed, as carefully integrated interior lighting sets the stone façade of the office floors aglow from within. In addition to creating a singular identity for his client (and notably doing so through architectural means rather than applied signage), Kuma aims to engage larger issues related to our perceptions and expectations of architectural materials, calling his design "a search into the ambiguous domain between two opposite items: the 'wall' and the 'window'" as well as the "ambiguous domain between 'stone,' the real, and 'stone image,' the virtual."[10] These provocative statements become clear as one understands the material selections for the building skin.

Kuma's dichotomy of wall and window is one with a long history in modern architecture. Responding to the development of the non-structural curtain wall, Arthur Korn wrote in 1929 that "it is now possible to have an independent wall of glass . . . no

longer a solid wall with windows . . . this window is the wall itself, or in other words, this wall is itself the window."[11] Yet Kuma is not so much interested in the elimination of the wall, or its replacement by all-glass curtain walls, as he is in a complex conflation of wall and window; it is a "both–and" proposition rather than "either–or." The traditional arrangement of stone wall containing punched window is both referenced and superseded by the LVMH façade.

Working with the façade consultants of New York-based Front Inc., Kuma developed a custom curtain-wall system with mullions spanning vertically from each floor to the next and supporting infill panels, as is common in multi-story office buildings. It is in the materials selected to form the curtain wall, however, that the system is innovative. The main vertical mullions are custom-built steel I-shapes, and they support translucent panels that each measure 0.91 meters wide by 4 meters high (3 feet × 13 feet-2 inches). The typical panel is a composite of a 4-millimeter thick (5/32-inch) onyx slab laminated with clear polyurethane resin between two panes of transparent float glass, the outer pane being 12 millimeters

5.19 LVMH Building, Osaka, Japan. Kengo Kuma and Associates, 2004.

5.20 Exterior view of illuminated curtain wall. LVMH Building, Osaka, Japan. Kengo Kuma and Associates, 2004.

5.21 Interior view of curtain wall. LVMH Building, Osaka, Japan. Kengo Kuma and Associates, 2004.

thick (½ inch) and the inner 6 millimeters (¼ inch), for a total panel thickness of approximately 25 millimeters (1 inch). The glass protects the more delicate onyx from wind and precipitation, and it facilitates cleaning of the façade. As custom-built components, these panels were subjected to a series of visual, seismic, and impact tests to confirm adequate performance prior to installation.

The onyx, quarried in Pakistan, is richly veined in hues ranging from beige and gold to a deep violet. Due to the transparent glass cladding on each side, the unique textures and patterns of each onyx slab are clearly visible from both inside and out. The onyx is thin enough to allow light to pass through, providing diffuse natural lighting to interior spaces and creating the desired glowing effect at night. As it is truly translucent, the onyx does not, therefore, allow viewing through the panels. Given the desire for workers inside the office spaces to have outside views, the architects incorporated a second type of panel into the curtain wall, one that omits the onyx slab, replacing it with a sheet of polyethylene terephthalate (PET) laminated between two glass panes. Each PET interlayer is primarily transparent but is also custom-printed with a pattern that simulates the grain of onyx. The laminated PET panels are interspersed with the onyx panels throughout the façade in a repeating pattern, with a ratio of one to two respectively, providing clear views to the exterior at every third panel (and as the two panel types are not identical in appearance, this pattern is discernible from the exterior under certain lighting conditions). In this way, Kuma plays with conceptions of the real and the virtual, and one's perception of the building shifts with the diurnal cycle.

The LVMH curtain wall contains another innovation which heightens the building's nighttime effect: each mullion incorporates an integral light fixture to illuminate the translucent façade directly. Instead of relying simply on ceiling lights or other interior task lighting to provide the glowing effect at night, this system provides dedicated lighting immediately behind the façade panels to supply consistent nighttime illumination. The primary purpose of a curtain-wall mullion is structural—it supports the infill panels and transfers loads to the building structural system beyond. At LVMH, the mullions perform this task efficiently,

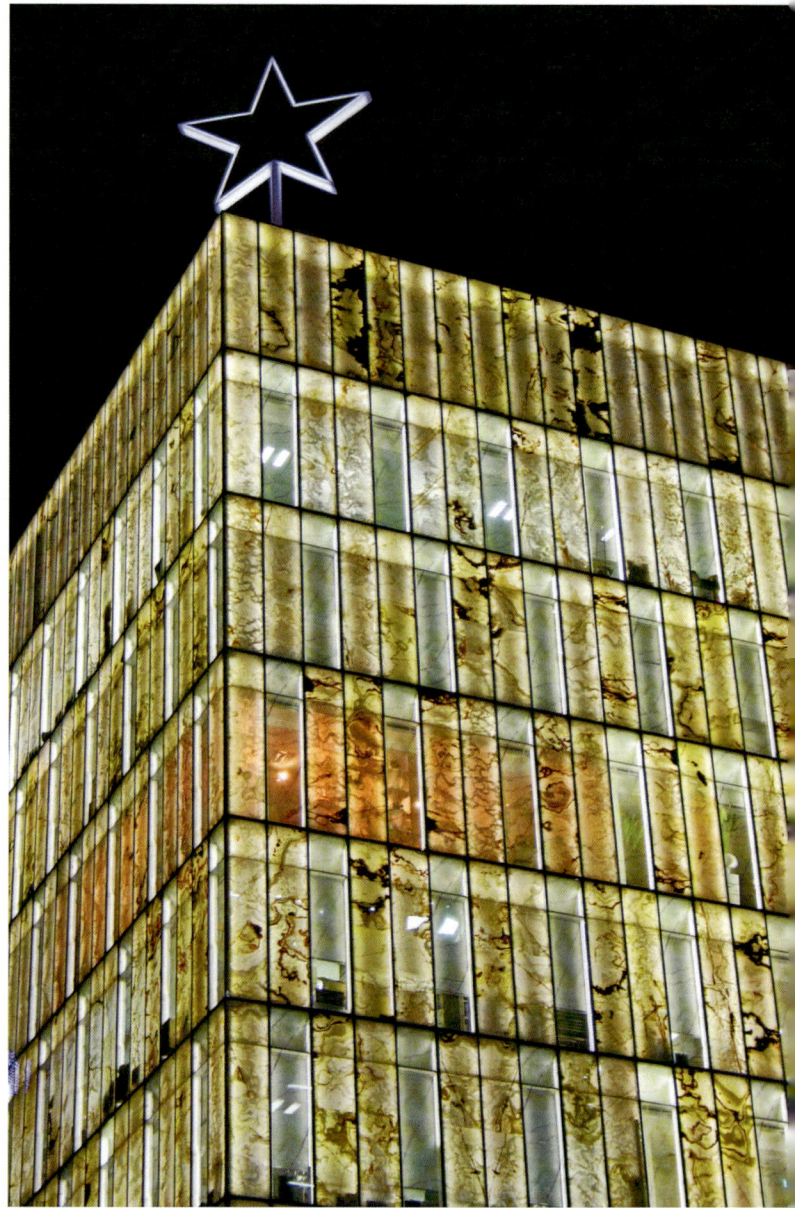

5.22 LVMH Building, Osaka, Japan. Kengo Kuma and Associates, 2004.

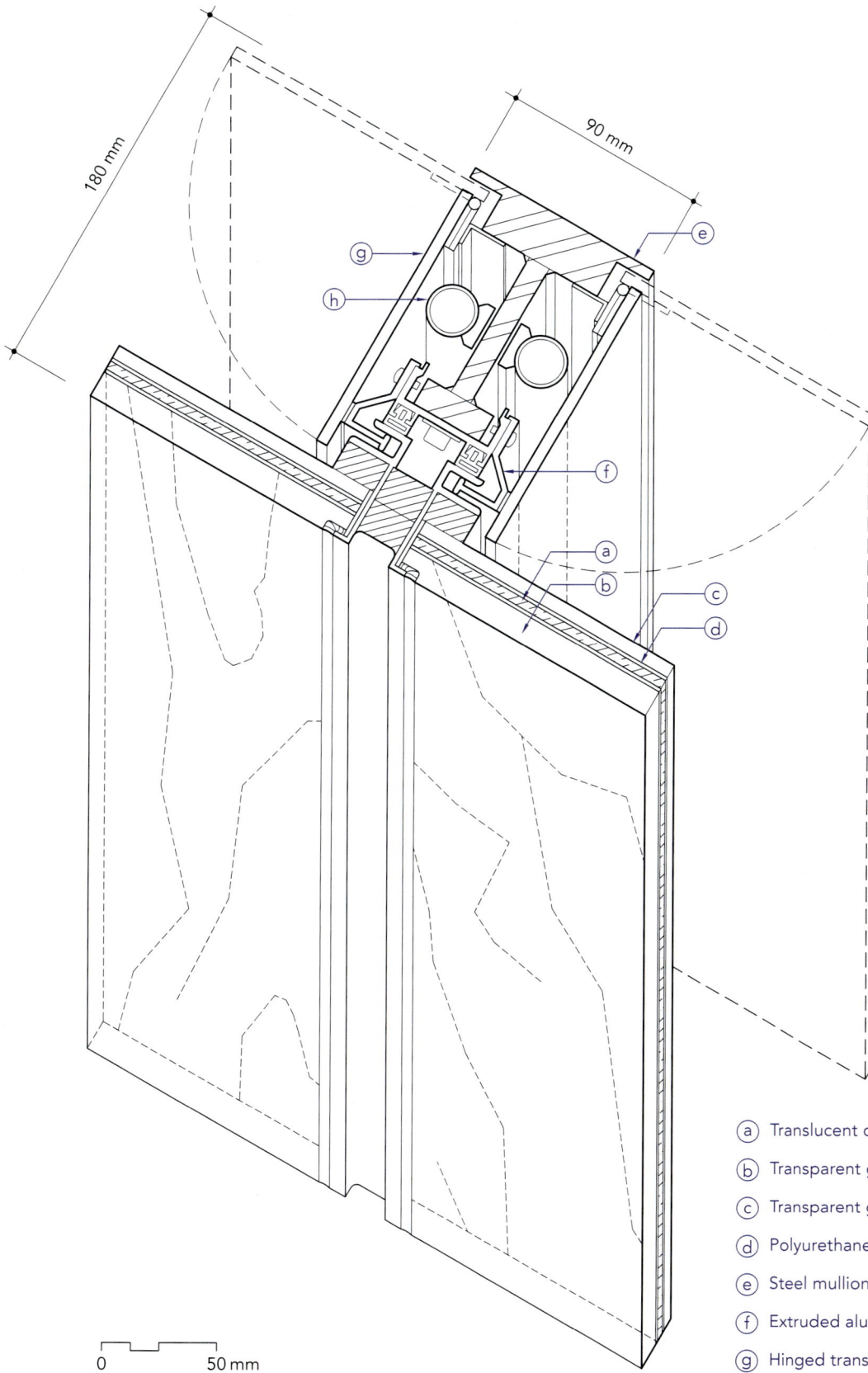

180 mm

90 mm

(g)

(h)

(e)

(f)

(a)

(b)

(c)

(d)

(a) Translucent onyx, 4 mm

(b) Transparent glass, 12 mm

(c) Transparent glass, 6 mm

(d) Polyurethane resin

(e) Steel mullion

(f) Extruded aluminum glazing adapter

(g) Hinged translucent glass cover, 5 mm

(h) Fluorescent light fixture

0   50 mm

5.23 Curtain-wall mullion detail. LVMH Building, Osaka.

5.24   LVMH Building, Osaka, Japan. Kengo Kuma and Associates, 2004.

utilizing an I-shape built of steel members. The glass and onyx panels were pre-glazed in a factory to small extruded aluminum adapter frames which were then connected on-site to the steel members. The glass-composite panels sit at the exterior face of the mullion, creating a flush surface to the outside and exposing the mullion to the interior. The space around the steel mullion (within the voids created by the I shape) is used to house linear fluorescent tubes, which are encased behind translucent glass panels that form the sides of the mullions and hide the light tubes from view. These side panels are hinged, allowing easy access for

maintenance. It is an ingenious system which results in three distinct modes of operation: during the day, the façade appears primarily opaque; at dusk, the interior volume becomes legible as office lights are turned on; at night, all of the mullion light fixtures are illuminated, creating the distinctive glowing lightbox effect.

Kengo Kuma has written that "materials only begin to show their true nature when you challenge them at the limits of their capabilities."[12] Kuma is well known for an approach to materiality that is both sensitive and experimental, and LVMH Osaka is just one of the numerous buildings he has designed with envelope

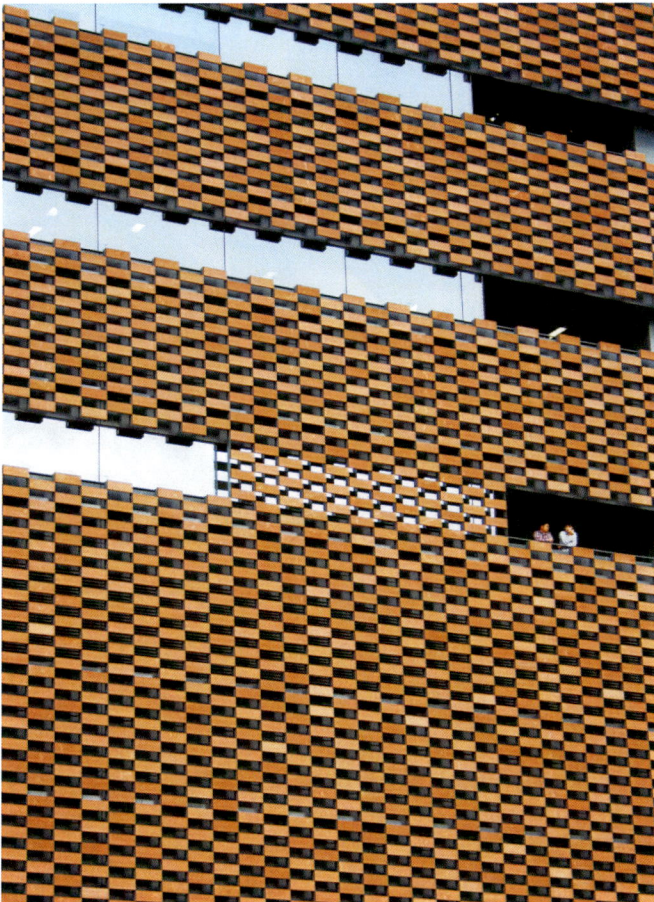

5.25 Asahi Broadcasting Corporation Headquarters, Osaka, Japan. Kengo Kuma and Associates, 2008.

5.26 Tiffany Flagship Store, Ginza, Tokyo, Japan. Kengo Kuma and Associates, 2008.

systems that deal with reflectivity, opacity, and translucency in interesting ways. In several of Kuma's projects, there is a clear strategy of developing layered façade systems to provide sunshading while also creating unique patterns and textures. Examples of this include the checkerboard-patterned composite-wood-panel sunscreens at the Asahi Broadcasting Corporation Headquarters (2008) in Osaka and the vertical granite louvers at the Nagasaki Prefectural Art Museum (2005). At the Tiffany Flagship Store (2008) in Tokyo's Ginza district, Kuma designed a

curtain wall with faceted planes of glass laminated to aluminum-honeycomb panels positioned at different angles to create subtle depth and reflectivity. A tinted-glass curtain wall is used at the Opposite House Hotel (2008) in Beijing, China, to develop a colorful pattern intended to reference traditional Chinese lattice screens.[13]

As described above, other architects have also explored the potential translucency of stone as a façade material. Architect Mariano Bayón designed an elegant veil of translucent marble for

5.27 Deutsche Bundesbank Headquarters, Chemnitz, Germany. Josep Lluis Mateo, 2004.

the Red Eléctrica Building (1992) in Seville, Spain. Steel-framed marble slabs suspended from columns do not enclose interior space, but create screen walls surrounding an open courtyard. The marble is illuminated at night, revealing the colorful veining of the stone panels, to become a symbol of the electrical company for which it was built.[14] Another project that should be noted is the Deutsche Bundesbank Headquarters (2004) in Chemnitz, Germany, by architect Josep Lluís Mateo. Here the main façade is a structural glass system consisting of insulating glass units that incorporate translucent panels of 15-millimeter thick (⅝ inch) alabaster laminated to tempered glass. The panels are bolted to spider fittings that are in turn suspended on vertical steel cables, which provide the necessary structural support for lateral as well as gravity loads. The building skin acts similarly to LVMH in that it provides ample diffuse daylighting for interior office and lobby spaces and transforms into a glowing lantern at night. Whereas Kuma conceived LVMH as a luminous stone box, Mateo likens the Bundesbank building to a block of ice or a "fossil that reveals its previous organic structure" through illumination of its organic patterning and veining.[15]

The two main projects discussed in this chapter—Beinecke Library and LVMH Osaka—represent two important protagonists in the ongoing narrative surrounding the roles of stone and glass in architecture. Bienecke Library is essentially a glass building housed within a stone building, creating an interstitial space of occupation. At LVMH Osaka, the stone and glass planes converge into a single surface. Both buildings derive the luminous qualities typically associated with glass through the unexpected vehicle of stone. Additionally, each building dramatically exploits the last century of evolution in stone-cutting and fabrication technology toward ever thinner slices of stone, such that a radical transformation in the concept of stone is possible: it can be thought of as a thin, translucent membrane rather than a massive, opaque block.

6
Bioluminescence

BETH SHOLOM SYNAGOGUE, ELKINS PARK, PENNSYLVANIA, USA, 1959

KURSAAL CONGRESS CENTER, SAN SEBASTIÁN, SPAIN, 1999

6.1   Kursaal Congress Center, San Sebastián, Spain. Rafael Moneo, 1999.

6.2 View from Playa Zurriola, day. Kursaal Congress Center, San Sebastián, Spain. Rafael Moneo, 1999.

The Kursaal Congress Center, designed by Rafael Moneo, occupies a dramatic site in the city of San Sebastián on the north coast of Spain.[1] The building is located at the mouth of the Urumea River, precisely at the boundary between a dense urban condition and the open sea—between city and nature. The design of the Kursaal was developed both in response to this unique situation and as an intentional remaking of it. In describing this project, Moneo writes that "architecture discovers the site, reveals it and makes it evident" and, concurrently, that "the site is where the specific object—the building—acquires its identity and finds its dimension, its unique, unrepeatable condition."[2] In this case, the architecture expresses its identity through two large, slightly tilted translucent glass volumes, distinct yet linked, that Moneo compares to two enormous rocks left stranded on the beach by the movement of the river.[3]

The Kursaal Center, completed in 1999, is a venue for concerts and conferences that is designed to accommodate a wide variety of events, from full-scale symphonic and opera performances to film festivals, lectures, and meetings. The two prismatic volumes described above contain the main programmatic components of the building, namely two auditoriums—a larger one to the west, with 1,806 seats, and a smaller one to the east, with 624 seats. Although in describing the project Moneo has often emphasized the building as an element of the landscape, removed from the city, the Kursaal does not ignore its urban responsibilities—it

respects the established building height of the adjacent commercial and residential districts, for instance, and offers an active zone of street-front shops and restaurants along a broad one-story podium that extends for two full city blocks.

The dual auditoriums are each expressed as a freestanding wood-clad mass encased by and floating within a secondary shell of glass, reaching heights of 30 meters and 24 meters (100 feet and 79 feet) respectively. The interstitial spaces created between each auditorium and its glass case become soaring canyons of circulation combined with open platforms and spaces for public gathering and lounging at various levels, all lit brilliantly in the daytime by natural light filtered through the translucent walls. At night, these same glass walls are illuminated from within, forming glowing beacons that operate at the scale of the city, as well as the landscape, and mark this site as a center of activity and public life. The building thus presents what can be termed an architectural occurrence of *bioluminescence*, defined as the phenomenon of light production and emission by a living organism.

The architectural component that enables this bioluminescence is the Kursaal's translucent building envelope, which encloses each auditorium and is in fact not a thin shell but a double-skin wall system with an overall depth of 2.4 meters (8 feet). The outer and inner skins each consist of translucent laminated glass panels, curved on the outer wall and flat on the inner. Each glass panel is 250 centimeters long (8.2 feet) and 60 cen-

6.3   View from Playa Zurriola, night. Kursaal Congress Center, San Sebastián, Spain. Rafael Moneo, 1999.

timeters high (2 feet), supported along its top and bottom edges by mullions of extruded aluminum. The concave panels of the outer skin are each curved to a radius of 60 centimeters (2 feet) and are comprised of two types of glass—a sheet of 19-millimeter thick (¾ inch) low-iron glass, sandblasted on one surface and laminated with a clear polyvinyl-butyral (PVB) interlayer to a sheet of 4-millimeter thick (³⁄₁₆ inch) fluted glass.[4] These glass types and fabrication processes were chosen for their specific properties: low-iron glass was selected for its clarity and absence of the slight greenish tint that is inherent to regular clear glass; sandblasting renders the glass translucent; fluted glass is embossed with a ribbed texture that emphasizes the horizontality of the glass and reflects light in unique ways. The flat glass panels of the inner skin consist of two sheets of 6-millimeter thick (¼ inch) sandblasted low-iron glass laminated together with a PVB interlayer. On the exterior, horizontal joints between glass panels are formed by custom cast-aluminum pieces while on the interior by aluminum caps with cedar covers. Vertical joints on both walls are de-emphasized, sealed simply with silicone. The resulting emphasis on horizontal joints, as well as the horizontal orientation of curved glass panels, suggests geological stratification and supports Moneo's rock metaphor. The space between the two glass walls contains the steel support structure to which the building envelope is anchored, as well as a series of steel-grate walkways that provide access for maintenance and cleaning of the façade.

As an entirely custom-designed system utilizing unique materials in non-standard assemblies, the building skin of the Kursaal required innovation in the fabrication processes used for its realization. The 5,000 curved glass units needed for the project were produced by the Barcelona-based glass fabricator Cricursa, which faced the challenge of bending two glass sheets of different thicknesses and then laminating them together, all within the tight tolerances and fast-tracked production schedule required by this high-profile building. The solution devised to accomplish this involved bending the two sheets simultaneously (rather than individually) to ensure matching curvature and continuous contact between the sheets during lamination. The process began by cutting flat glass sheets to size, polishing the edges, and then performing the sandblasting process. Then, in order to ensure that the two pieces of curved glass that comprise each panel would precisely fit together for lamination, they were first stacked together as flat pieces on top of a curved steel formwork and then heated until softened, at which point the glass would slump into the form, taking on the specified curvature. The glass pieces were then cooled, laminated together with a PVB interlayer, and delivered to the construction site for installation.[5]

These complex combinations of glass treatments and forms are unique to the Kursaal, and, through the interaction of these assemblies with light, result in similarly complex and shifting effects on the building's experiential qualities. Because of its

6.4 Concave translucent laminated glass. Kursaal Congress Center, San Sebastián, Spain. Rafael Moneo, 1999.

placement, the Kursaal is experienced on one side (the south) as part of the city but on the other (the north) as an element of the landscape. From each perspective, the building skin's character can change dramatically with environmental conditions. Under cloudy or overcast skies, the walls can appear opaque, like stone, or iridescent. Direct sunlight, however, is able to partially penetrate the translucent outer skin to reveal the grid of structural steel elements encased within the double-skin cavity. The horizontally oriented scalloped-glass panels create an alternating pattern of shade and highlights that corresponds to the angle of the sun. At

midday, the glass skin usually appears slightly greenish or bluish in tone, but at sunrise and sunset reflects the brilliant oranges and reds of the sky. At night, the Kursaal becomes an enormous lantern, as light fixtures placed within the double-skin cavity create the building's distinctive radiance. Inside the building, the flat translucent glass walls provide remarkably consistent lighting to the lobby and circulation spaces that surround each auditorium, revealing the silhouette of hidden structure and giving the building's interior an identity as strong as its exterior.

There is an intentional bifurcation of experience between

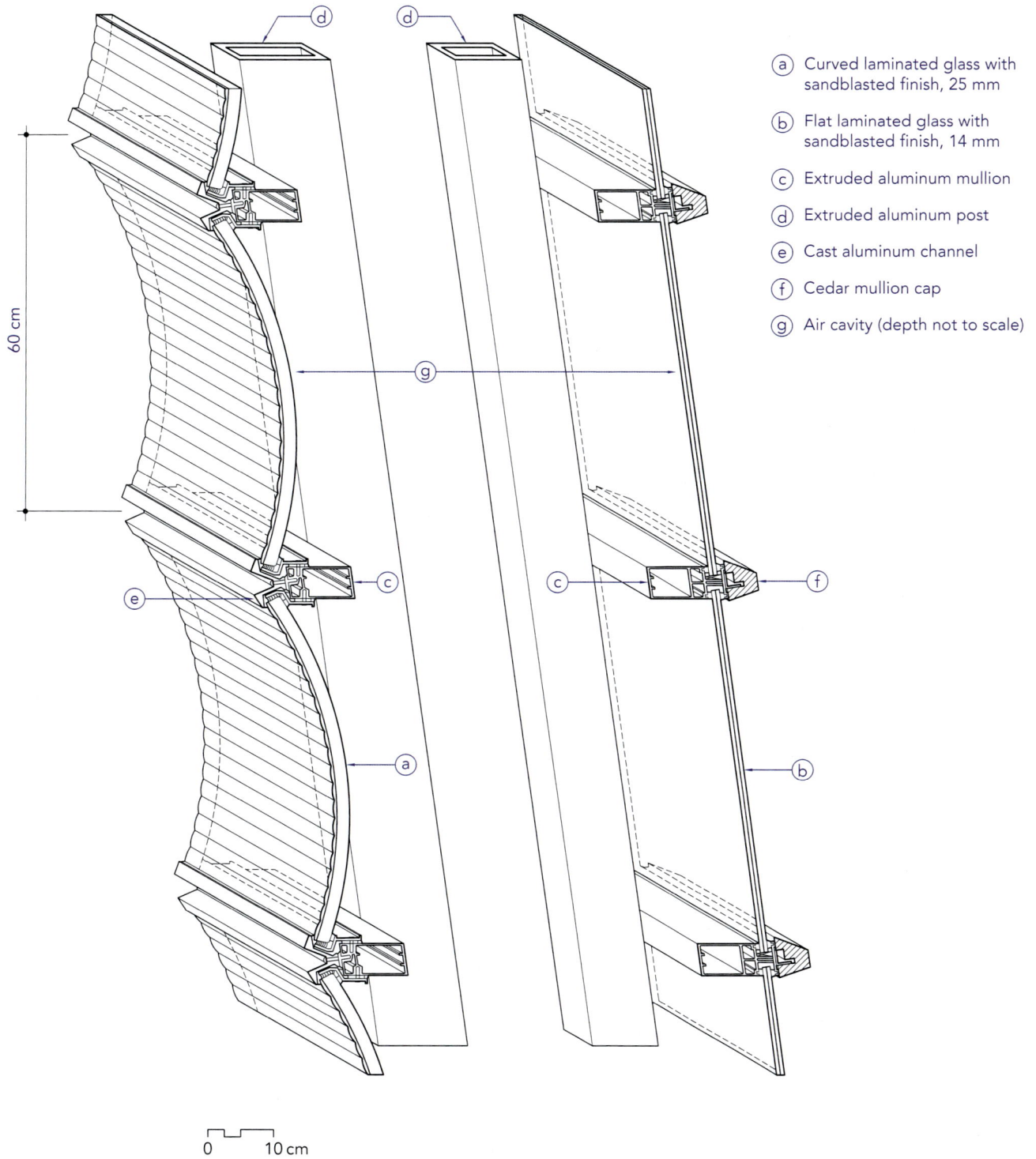

(a) Curved laminated glass with sandblasted finish, 25 mm

(b) Flat laminated glass with sandblasted finish, 14 mm

(c) Extruded aluminum mullion

(d) Extruded aluminum post

(e) Cast aluminum channel

(f) Cedar mullion cap

(g) Air cavity (depth not to scale)

60 cm

0   10 cm

6.5   Double-skin curtain wall. Kursaal Congress Center.

6.6 View from Playa Zurriola, night. Kursaal Congress Center, San Sebastián, Spain. Rafael Moneo, 1999.

6.7 Kursaal Congress Center, San Sebastián, Spain. Rafael Moneo, 1999.

6.8 Interior. Kursaal Congress Center, San Sebastián, Spain. Rafael Moneo, 1999.

6.9 Interior, framed view toward Playa Zurriola. Kursaal Congress Center, San Sebastián, Spain. Rafael Moneo, 1999.

6.10 Interior. Kursaal Congress Center, San Sebastián, Spain. Rafael Moneo, 1999.

6.11 55 Water Street Plaza, New York City, USA. Rogers Marvel Architects and Ken Smith Landscape Architect, 2005.

6.12 Dewey Square T-Station, Boston, Massachusetts, USA. Machado and Silvetti Associates, 2007.

inside and outside. Despite the Kursaal's immediate proximity to the picturesque Zurriola Beach and the Bay of Biscay, Moneo has deliberately restricted vision outward from the building, providing only a relatively few framed views of the sea through strategically placed transparent windows. One's experience within the building therefore remains focused on the interior environment, where Moneo intends the translucent skin to provide an allegorical rather than literal engagement with water, what he has called "a very different experience, to do with humidity, with wetness and water —almost the feeling of going under the sea."[6]

The Kursaal is certainly unique among the buildings of San Sebastián and, through its distinctive materiality and luminosity, acts as an urban beacon, drawing attention to its prime location and to the events and activities it supports. An analogous effect can be seen in two recent projects in the USA which are much smaller in scale but take a similar approach to materiality. The 55 Water Street Plaza is an elevated public space set among the skyscrapers of lower Manhattan's riverfront and was designed by Rogers Marvel Architects in association with Ken Smith Landscape Architect.[7] Completed in 2005, this one-acre mini-park includes informal seating areas, lawns, plantings, and an amphitheater. The most visible element of the design, however, is a pavilion standing 15 meters (50 feet) tall and built of translucent glass panels mounted to a structural steel frame. The glass is illuminated by colored light-emitting diodes (LEDs) integrated into the framing. At night the glowing pavilion is visible from as far away as Brooklyn Heights, across the East River. Officially named the Beacon of Progress, it is a contemporary lighthouse that recalls an older structure, the Titanic Memorial Lighthouse, which occupied the same site from 1913 to 1968.[8] In 2007, architects Machado and Silvetti completed a redesign of Dewey Square, in Boston's financial district, that includes new head-house entrances to the subway below. The steel-framed head houses are clad entirely in translucent glass panels and are designed, in the architects'

6.13 Kursaal Congress Center, San Sebastián, Spain. Rafael Moneo, 1999.

words, to "give the precinct a distinctive identity and serve as glowing beacons at night."[9] More recently, and with a program and scale quite similar to the Kursaal, these ideas are engaged by the Grand Theatre (2009) in Chongqing, China, designed by German architects von Gerkan, Marg & Partner. This building is sited along the Yangtze River and contains two theaters wrapped by a double-skin curtain wall of translucent glass. With its prominent location and dramatic nighttime illumination, it has quickly become a landmark in the city of Chongqing.

While glass has been commonly perceived by the modern architect as an "invisible" membrane, its transparency enabling a conceptual continuity between interior and exterior, the Kursaal Center and other buildings like it mark an emphatic break with this tradition. They present an alternative conception of glass as present rather than absent, tactile rather than visual, thick rather than thin, and ambiguous rather than obvious. When glass is transformed from transparent to translucent, it acquires the capacity to capture and embody light, rather than simply transmitting it, lending these buildings their unusual presence.

* * *

The physical embodiment of light takes on religious significance and a defining role at Frank Lloyd Wright's Beth Sholom Synagogue, built in 1959 in Elkins Park, a suburb north of Philadelphia, Pennsylvania.[10] Hexagonal in plan, the low concrete walls of the building's ground floor support a towering pyramidal structure above, rising to a height of 33.5 meters (110 feet).[11] This tent-like translucent glass roof floods the synagogue's interior space with natural light by day and emits a silvery glow at night.

Wright's design is rich with metaphor and symbolism, developed in collaboration with his client, Rabbi Mortimer J. Cohen, who initially led the synagogue in seeking out Wright for the project and was intimately involved throughout the design and construction phases.[12] Architect and client were united in the goal of creating an unapologetically modern incarnation of Judaism.[13] The form of the synagogue is meant to abstractly represent Mount Sinai, the sacred flat-topped mountain where Moses received the Torah from God. At that time, according to Jewish tradition,

6.14 Beth Sholom Synagogue, Elkins Park, Pennsylvania, USA. Frank Lloyd Wright, 1958.

6.15   Beth Sholom Synagogue, Elkins Park, Pennsylvania, USA. Frank Lloyd Wright, 1958.

Mount Sinai was ablaze with light, a condition signified by Beth Sholom's radiant glass roof. Along the top edge of each of the three main roof beams are placed seven lamps, intended as stylized versions of the Menorah visible from all sides, which shine upward to illuminate the structure at night. The interior space, with its luminous white ceiling rising to a central point, is meant to recall the tabernacle tents of ancient nomadic worshippers. Wright also claimed that the irregular hexagon of the floor plan resembled the two cupped hands of God, held together to support the congregation.[14]

Beth Sholom can certainly be considered a mature work by Wright, who was 86 years old when he was hired in 1953 to design the synagogue and died just five months before its official opening in 1959. It is also unique: during his long career, Wright designed more than 1,100 projects (about 530 of which were constructed), but Beth Sholom was his only commission for a synagogue. The design clearly draws upon Wright's earlier unbuilt plan for the "Steel Cathedral for a Million People," which he had proposed for New York City in 1926, and is also loosely reminiscent of the luminous canvas roofs of Taliesin West, Wright's winter home built in 1937 in the Arizona desert. Furthermore, the importance of glass and light to the modern condition—a formative concept for his synagogue design—had been a preoccupation of Wright's for decades. In the second of his six Kahn Lectures (a series titled simply "Modern Architecture") at Princeton University in 1930, Wright appears, in retrospect, to forecast by 25 years his conceptual and material strategies for Beth Sholom:

Shadows were the "brush-work" of the ancient Architect. Let the "Modern" now work with light, light diffused, light reflected, light refracted—light for its own sake, shadows gratuitous. It is the Machine that makes *modern* these rare new opportunities in Glass . . . The Prism has always fascinated man. We may now live in prismatic buildings, clean, beautiful and New. Here is one clear "material" proof of modern advantage, for Glass is uncompromisingly Modern. Yes—Architecture is soon to live anew because of Glass and Steel.[15]

6.16 Interior. Beth Sholom Synagogue, Elkins Park, Pennsylvania, USA. Frank Lloyd Wright, 1959.

Beth Sholom's prismatic surfaces of steel-framed glass, which may be read as either inclined walls or faceted roof planes, enclose the synagogue's main sanctuary. This column-free space seats 1,028 people and is located on the second level, accessed from the low-ceilinged ground-floor lobby by a pair of opposing staircases that flank the main entrance doors. The floor of the sanctuary is gently inclined downward to the east, toward the pulpit, while the space soars dramatically overhead. Large steel beams, encased in concrete, rise from three corners of the room like a giant tripod, meeting high above the center of the space, where a colorful Wright-design chandelier of stained glass hangs. Each primary beam is 36 meters (117 feet) in length. A secondary network of horizontal and vertical steel members spans between the main beams, forming the sloped planes of the roof and supporting the translucent panels of the enclosure system. The steel elements are clad with cast aluminum covers, some of which are stamped with an abstract decorative motif characteristic of Wright's work from this period.

The main building skin is comprised of two layers: an outer one consisting of sandblasted white corrugated wired-glass panels arranged in a shingle configuration, and an inner layer of corrugated glass-fiber-reinforced plastic.[16] The two layers are separated by an insulating air cavity several centimeters in depth. The typical outer panels are each approximately 70 centimeters wide by 140 centimeters high (28 by 55 inches) while the inner ones are the same height but slightly wider. In the early stages of the design, Wright had planned to use a blue-tinted plastic panel for the inner layer of the double skin but later changed it to a more neutral ivory-colored fiberglass, probably in recognition that light passing through the tinted plastic would have given the interior space, and those in it, a distracting and likely unflattering bluish hue.[17] Panel joints are expressed differently for each layer: on the exterior surface, horizontal joints are formed by the overlapping edges of the glass panels while vertical joints are covered with aluminum strips; on the interior, horizontal joints are emphasized with protruding stamped aluminum-clad mullions while vertical joints are minimally covered with thin aluminum straps. This offset arrangement of mullions results in a pattern of opposing shadows cast on each surface that indirectly reveals the depth of

6.17 Beth Sholom Synagogue, Elkins Park, Pennsylvania, USA. Frank Lloyd Wright, 1959.

6.18 Sanctuary interior. Beth Sholom Synagogue, Elkins Park, Pennsylvania, USA. Frank Lloyd Wright, 1959.

6.19 Beth Sholom Synagogue, Elkins Park, Pennsylvania, USA. Frank Lloyd Wright, 1959.

the skin and the presence of the other layer. The glazing system of shingled glass panels recalls techniques traditionally used on greenhouses since the nineteenth century but which are adapted and embellished through Wright's material choices and framing details.

The sanctuary at Beth Sholom has virtually no opaque walls or transparent glass. It is permeated with a remarkably consistent and diffuse light transmitted by its translucent enclosure. This space, with its inwardly focused, meditative character, must have been an influence on James Stirling's design of a decade later for the History Faculty Building (1967) at Cambridge University. In Stirling's building, a great reading room six-stories high is capped by sloping planes of translucent glass arranged in a tent-like fashion similar to Beth Sholom. The roof above the reading room consists of structural steel trusses clad with upper and lower skins of glass, with a ventilated interstitial space as high as 3.6 meters (12 feet) in places. The lower skin is translucent glass, intended to provide diffuse, shadowless natural light to the reading tables below.[18] The serene qualities of the Beth Sholom Synagogue can also be found in later buildings constructed by people of various religious faiths, evoking the universal association of light with divine presence. A recent example is the Prayer Pavilion of Light (2007) in Phoenix, Arizona, designed by DeBartolo Architects. This 232-square meter (2,500-square foot) glass-box chapel was commissioned by the Pentecostal Phoenix First Assembly of God and provides a space for meditation that is open to the public 24 hours a day, seven days a week. Here the primary space is encased by double-skin walls consisting of an outer layer of laminated glass and an inner layer of triple-pane insulating glass units, supported by steel Vierendeel trusses. Between the skins is a 1.5-meter deep (5 foot) interstitial cavity, vented at the top and bottom to alleviate heat gain. The glass in each layer is rendered translucent through the application of a ceramic frit coating designed to simulate the appearance of sandblasted glass (but which is much easier to clean). The lower portion of each wall incorporates sliding glass doors which can be moved aside to completely open the space to the exterior landscape and weather, leaving a kind of halo of glass hovering around the sanctuary. Like Beth Sholom, the Prayer Pavilion relies upon abundant natural lighting during the day

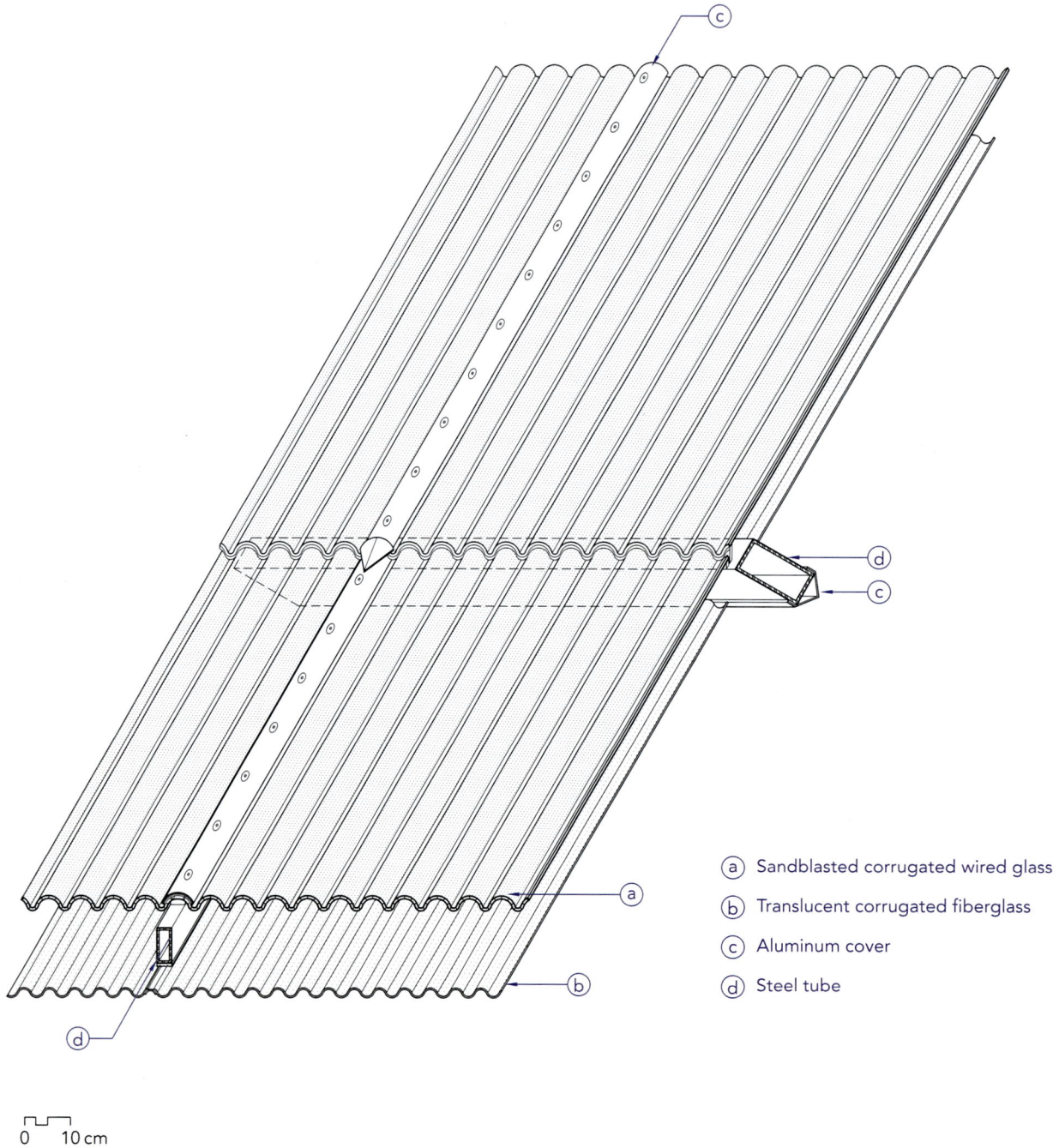

(a) Sandblasted corrugated wired glass

(b) Translucent corrugated fiberglass

(c) Aluminum cover

(d) Steel tube

0   10 cm

6.20 Double-skin envelope assembly. Beth Sholom Synagogue.

Interior view of double-skin envelope. Beth Sholom Synagogue, Elkins Park, Pennsylvania, USA. Frank Lloyd Wright, 1959.

6.22 Interior view of wall with integrated light fixture. Beth Sholom Synagogue, Elkins Park, Pennsylvania, USA. Frank Lloyd Wright, 1959.

6.23 Interior view of ceiling. Beth Sholom Synagogue, Elkins Park, Pennsylvania, USA. Frank Lloyd Wright, 1959.

6.24 History Faculty Building, Cambridge University, Cambridge, UK. James Stirling, 1967.

6.25 Prayer Pavilion of Light, Phoenix, Arizona, USA. DeBartolo Architects, 2007.

135

and becomes a lantern-like symbol at night. Its envelope cavity contains strips of LEDs that can shift in color, illuminating and animating the translucent glass skin at night, making this glowing cube visible from miles away.[19]

Though overshadowed by many of Wright's more widely appreciated buildings, including the Guggenheim Museum, which was under construction in New York City at the same time, the Beth Sholom Synagogue was eventually recognized as a building with historical significance—Brendan Gill called it "one of the most important events in Wright's career."[20] Within a few years of its completion, Beth Sholom was listed by the American Institute of Architects as one of the 17 Wright projects most worthy of preservation, and in 2007 it was named a US National Historic Landmark (the only twentieth-century synagogue so named). In addition to its representation of a particular culture and time, this building illustrates how spatial and material innovations can be combined with their resultant effects on light and atmosphere to inspire awe and bring to architecture a sense of the divine. At Beth Sholom, Wright and Cohen successfully achieved their original goal of providing a modern expression of traditional meaning, largely through the abstraction afforded by the conditions of translucency and luminescence.

6.26 (opposite) Interior. Beth Sholom Synagogue, Elkins Park, Pennsylvania, USA. Frank Lloyd Wright, 1959.

7

Fade to Black

7.1 InterActiveCorp Headquarters, New York City, USA. Gehry Partners, 2007.

The InterActiveCorp (IAC) Headquarters represents several firsts in the long career of its architect, Frank Gehry. Completed in 2007 in the Chelsea neighborhood of Manhattan's far west side, IAC is Gehry's first building in New York City (he was 78 years old at the time).[1] It is the first time Gehry has designed a building envelope consisting entirely of glass curtain wall. And it is the first building in which Gehry has so emphatically engaged with the condition of translucency—here experienced as an oscillation between opacity and transparency, between solid and void—as a primary element of the building skin.

Formally, the IAC Headquarters incorporates Gehry's characteristic language of fluidity: irregularly twisting, billowing, and torqued volumes call to mind marine imagery of waves, fish, or ships. Standing 49 meters (160 feet) in height, the building contains 12,100 square meters (130,000 square feet) of space on ten floors and was constructed at a cost of 100 million dollars. In addition to a ground-floor lobby and exhibition space and a top-floor café, the program primarily consists of offices and meeting spaces for IAC, an Internet company that oversees more than 50 website brands, including Ask.com, Match.com, and CollegeHumor. IAC chairman Barry Diller, who commissioned Gehry for the project, wanted a new headquarters to unite the company's numerous and diverse divisions under one roof for the first time. Diller said, "We were trying to do something new with this company, and I wanted a different kind of place—stimulating, beautiful, full of light."[2] Gehry's design delivers an assemblage of unique forms unified by a continuous wrapper of aluminum-framed glass with distinctively graphic horizontal banding, fading from opaque to translucent to transparent and back again at each level. The custom design of this glass skin appears to respond to dual goals: creating a recognizable and unifying identity for the building (without resorting to applied graphics or signage)[3] and addressing the technical performance requirements of the building envelope through non-traditional means. IAC likewise makes use of advanced digital tools for both design and fabrication and its construction required the invention of some innovative construction techniques.

Issues of materiality and their resultant experiential effects have constituted a long-running theme in the work of Frank Gehry and

7.2   InterActiveCorp Headquarters, New York City, USA. Gehry Partners, 2007.

his associates over more than four decades. In his early projects, Gehry often explored the potential of utilitarian materials to make architectural enclosures: he famously used chain-link fencing on the façade of his own house (1978) and at the Edgemar Development (1988), both in Santa Monica, California. Many of Gehry's mid-career projects combine complex form with a simple, almost minimalist material palette. His Vitra Design Museum (1989), in Weil am Rhein, Germany, for instance, uses white plaster walls and zinc roofing for its primarily opaque envelope. At the Vontz Center for Molecular Studies (1999) in Cincinnati, Ohio, the bowing walls are clad with brick (actually brick-faced prefabricated concrete panels) with large punched openings filled with transparent glass curtain walls. An earlier glimpse of an interest in translucency can be seen in Gehry's Center for the Visual Arts (1992) at the Toledo Museum of Art. The building is mainly clad with shingles of lead-coated copper but also includes some window walls with translucent sandblasted glass panels, each left with a rectangle of transparent glass in the center (similar to IAC)

to allow unimpeded views. Eventually Gehry turned to the use of metal as his cladding material of choice, given its pliability in sheet form and the relative ease of bending it to fit his building's curvaceous and faceted forms. Most famous among this group of projects is the Guggenheim Museum (1997) in Bilbao, Spain, covered primarily with its fish-scale shingles of titanium sheet. Stainless-steel panels are used to clad the undulating walls of the Walt Disney Concert Hall (2003) in Los Angeles, California, and the Beekman Tower (2011) at 8 Spruce Street in lower Manhattan.[4]

Early study models for the IAC Headquarters show that Gehry considered a range of cladding materials, including ones he had used extensively before, such as stone, stainless steel, and titanium panels, each combined with a repeating pattern of rectangular punched windows.[5] However, Barry Diller reportedly insisted on an all-glass building, perhaps in pursuit of novelty, since it was something Gehry had not done before, or perhaps in response to the dominant material palette of the new architecture sprouting up in the area (with Richard Meier's glass-box Perry

7.3 Vitra Design Museum, Weil am Rhein, Germany. Gehry Partners, 1989.

7.4 Center for the Visual Arts, Toledo Museum of Art, Ohio, USA. Gehry Partners, 1992.

7.5 Vontz Center for Molecular Studies, Cincinnati, Ohio, USA. Gehry Partners, 1999.

7.6 Walt Disney Concert Hall, Los Angeles, California, USA. Gehry Partners, 2003.

7.7   InterActiveCorp Headquarters, New York City, USA. Gehry Partners, 2007.

Street Apartments having set the tone in 2002). Having made the decision to make IAC a glass building, the design team was faced with the question of what type of glass to use. Given the restrictions of New York City's energy code and the inherently poor performance of glass in terms of thermal insulation and solar heat gain, a skin of fully transparent glass was not an option. One early model of the building included an envelope of mirror-reflective glass that would have successfully mitigated the solar-heat-gain issue. This option was rejected by Barry Diller for its association with cheap, generic office buildings of previous decades.[6] In the end, the designers chose to use double-pane insulating glass units with a low-e (low emittance) coating (to improve thermal performance) and an additional coating of a custom white-colored ceramic frit pattern that would partially reflect solar energy and provide shading while still allowing views into and out of the building where desired. Each of the two panes of glass is 10 millimeters thick (⅜ inch), with one laminated and one monolithic, separated by an air-space of 12 millimeters (½ inch). Ceramic frit— basically a ceramic-based paint that can be silkscreened onto glass and then baked to fuse the paint with the surface of the glass —can be applied as a full-coverage coating or in an intermittent

pattern (most often lines or dots) and is usually black or white, although custom colors can also be achieved. For the IAC curtain wall, the architects developed a pattern of white frit that transitions from full-coverage at the floor and ceiling of each level to a zone of gradated dots that gradually diminish in density until reaching a band of transparent glass roughly at eye level for a person standing inside. The dots are each about 1.5 millimeters (⅟₁₆ inch) in diameter and are only legible from within a meter or two of the glass. From greater distances, they blend together visually to give the appearance of varying degrees of translucency, like a solid fading into mist. The overall effect is neither opaque nor transparent, but occupies a hazy state in between.

In the last two decades, architects have been increasingly drawn to experimentation with glass fabrication techniques in general—and with ceramic frit patterning on glass in particular. The IAC façade therefore should be viewed as part of this spectrum of work. A veil-like effect quite similar to that of the IAC façade can be found at Norman Foster's B3 Office Building (1989) in Stockley Park, London, one of the earliest examples of ceramic frit patterning used primarily for solar control in a curtain-wall system but deployed in an artful manner. Glass with a solid

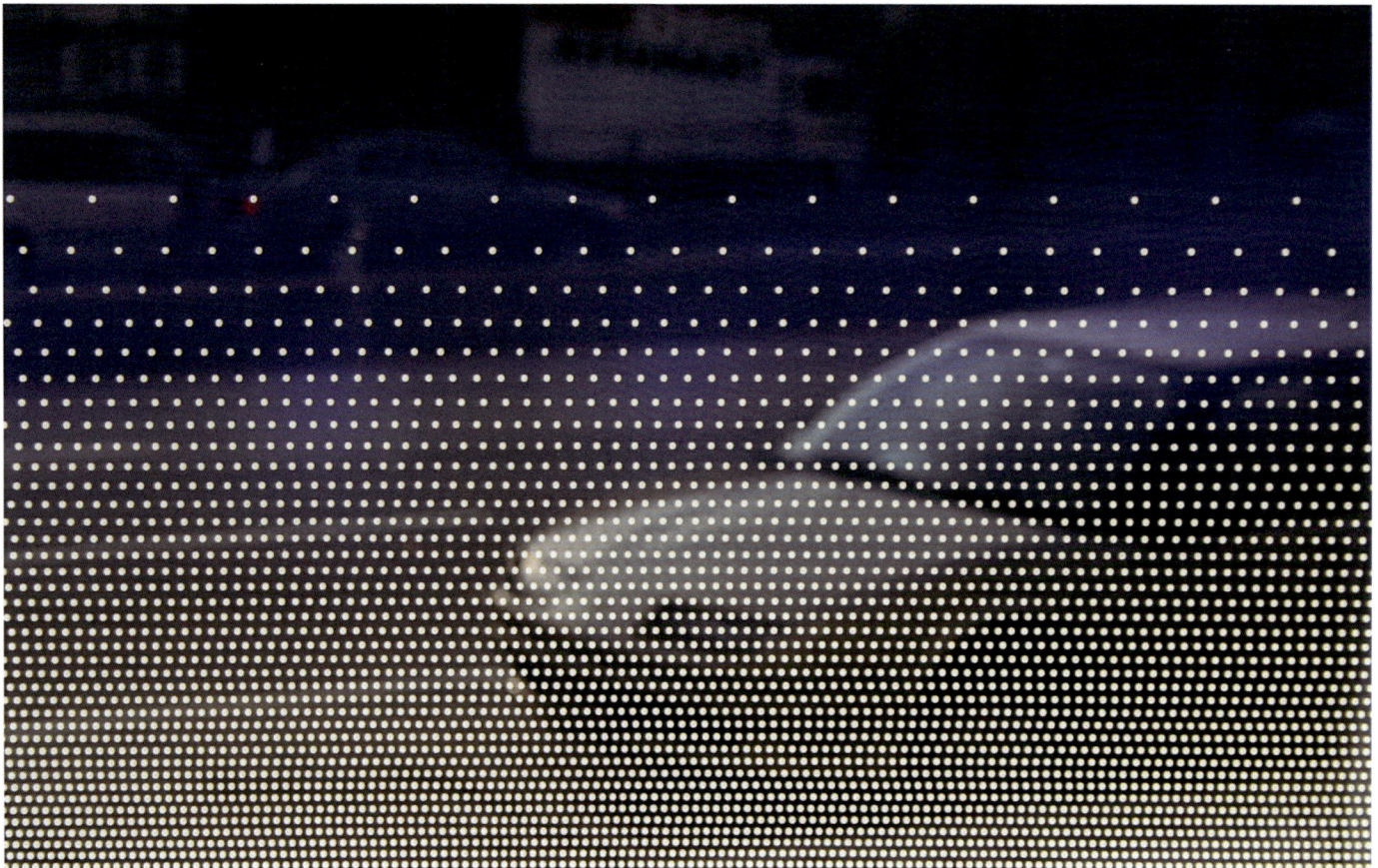

7.8 Close-up exterior view of ceramic frit dot pattern on curtain-wall glass. InterActiveCorp Headquarters, New York City, USA. Gehry Partners, 2007.

coating of white frit is used to conceal the floor plates, while vision areas between floor and ceiling consist of insulating glass panels with a white dot pattern that gradually decreases in density, leaving an eye-level band of nearly—though not completely—transparent glass in the middle. The gradual fade pattern at B3 is more successful in presenting a unified façade than the one at IAC, where the transparent zone is too wide and the fade too fast, resulting in a high-contrast expression of stripes. Atelier Christian de Portzamparc used patterned ceramic frit in their influential façade for the LVMH Tower (1999) in New York City as well as at the Hotel Renaissance (2009) in Paris, where frit lines create a fad-

ing pattern similar to IAC's. For a recladding of the Louis Vuitton Building (2004) in New York City, Jun Aoki devised a shifting pattern of white ceramic frit squares printed on different layers of a laminated glass assembly. In Chicago, the faceted glass curtain wall of the Spertus Institute of Jewish Studies (2007) by Krueck + Sexton Architects is silkscreened with a pattern of white ceramic frit dots covering 40 percent of the surface. Architects Jakob + Macfarlane created an undulating, translucent addition to an existing shipping depot at the Docks en Seine (2008) in Paris using a white ceramic frit dot pattern printed on green glass. Usually deployed in a manner meant to deemphasize the frit itself in favor

7.9 B3 Office Building, Stockley Park, London, UK. Norman Foster, 1989.

of a larger abstract effect, the technique has also been used to give façades a strikingly graphic impact through recognizable imagery and text. At the Cartier Administration Building (1990) in Fribourg, Switzerland, for example, Jean Nouvel designed a glass curtain wall imprinted with a repeating pattern of large-scale ceramic frit Cartier logos, essentially creating a building-integrated billboard. For the glass façade at the Utrecht University Library (2004), Wiel Arets designed a silkscreened figurative image of papyrus plants that renders the glass translucent and transmits diffuse natural light to the reading rooms within. And for the IKMZ Building (2004) at Cottbus University in Germany,

Herzog & de Meuron designed an all-glass envelope incorporating white custom-silkscreened patterns of overlapping handwritten texts, suggestive of accumulated layers of graffiti.

This discussion of the IAC skin has thus far focused on the fabrication of the glass itself, its composition and its visual and technical performance. The other important aspect of this particular building envelope is of course the geometrically complex form to which the glass curtain wall had to be molded. The fabrication of curved glass is a costly process, and to use curved glass for the IAC façade would have been prohibitively expensive, as nearly each piece of glass would need to be bent to a unique

7.10 Louis Vuitton Building, New York City, USA. Jun Aoki, 2004.

7.11 Spertus Institute of Jewish Studies, Chicago, Illinois, USA. Krueck + Sexton Architects, 2007.

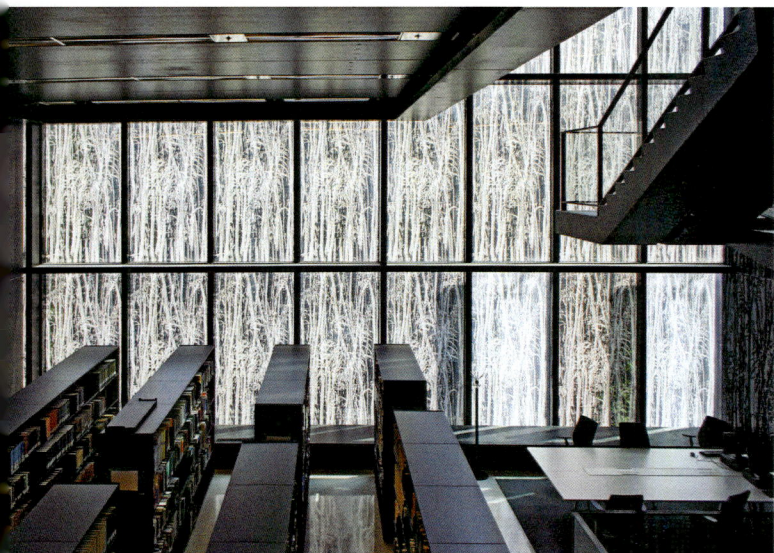

7.12 Utrecht University Library, the Netherlands. Wiel Arets Architects, 2004.

7.13 IKMZ Building, Cottbus University, Germany. Herzog & de Meuron, 2004.

150 cm

400 cm

0   1 m

(a) Laminated insulating glass unit with custom-patterned ceramic frit

(b) Extruded aluminum unit frame

(c) Reinforced concrete floor slab

(d) Adjustable curtain-wall anchor

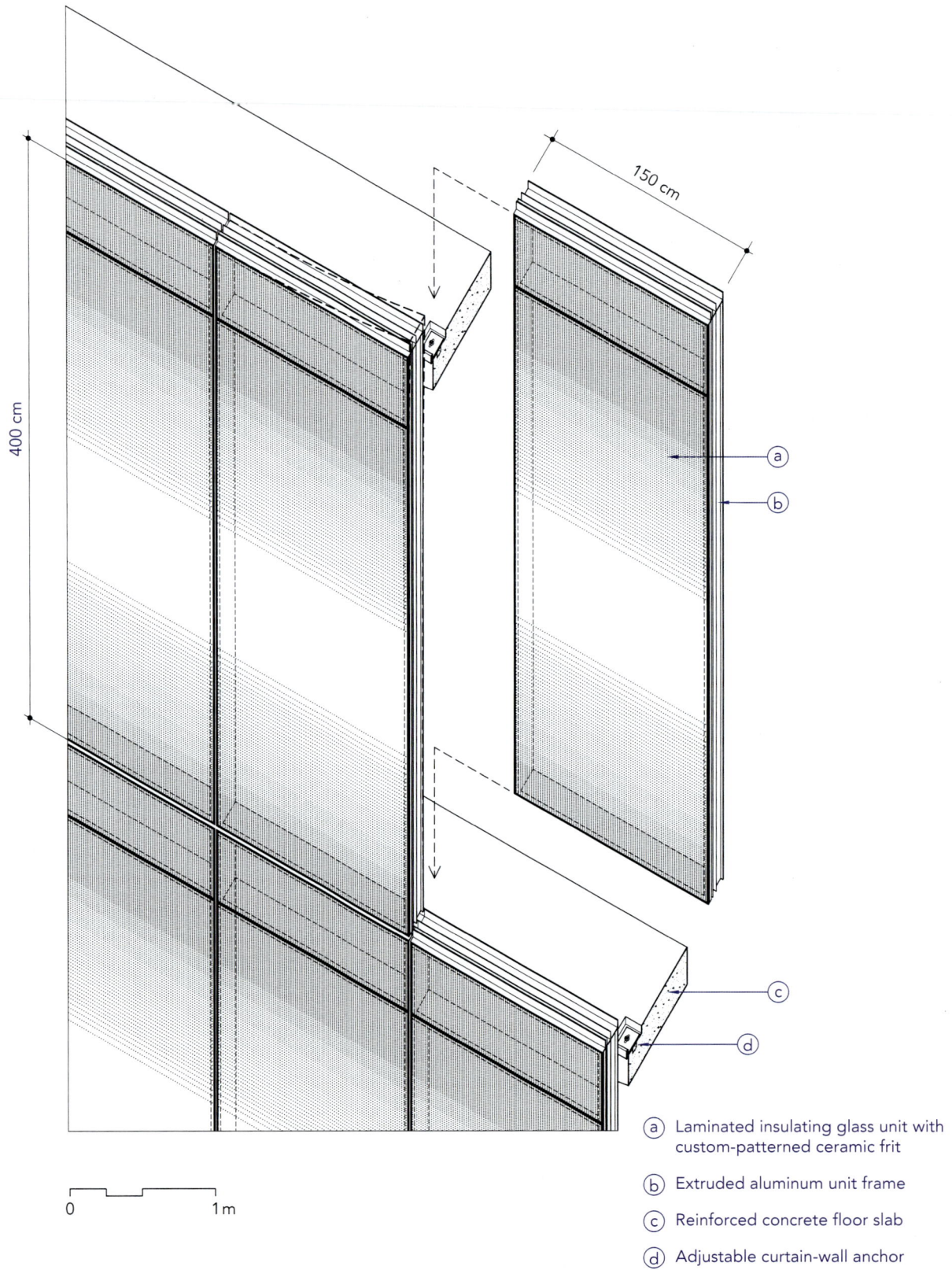

7.15 Unitized curtain-wall system. InterActiveCorp Headquarters.

7.16 Installation of curtain wall. InterActiveCorp Headquarters, New York City, USA. Gehry Partners, 2007.

radius in order to achieve the overall building form. Instead, the curtain-wall units were prefabricated in a factory as flat units, incorporating the extruded aluminum edge-frames and glass sheets, and were bent into the required warped shapes during their

installation at the construction site. This process is known as cold-bending or cold-warping (referencing the distinction that the glass is not bent in the typical manner—through application of heat to soften the glass) and was developed for IAC by Permasteelisa, a fabricator which had previously built other Gehry façades, including the Walt Disney Concert Hall.

This operation began with the architects' digital 3D building-information model, which contained dimensionally accurate representations of each façade component. Of the building's 1,450 total curtain-wall units, 1,150 are unique in shape. In general dimensions, each curtain-wall unit is about 1.5 meters wide (5 feet) by the height of one story, ranging from about 4 to 4.6 meters (13 to 15 feet). Permasteelisa, working with Gehry's office and their curtain wall consultant, Israel Berger and Associates, analyzed the degree of bending needed for each unit to confirm that it was within an acceptable range. The calculated limit of about 10 centimeters (4 inches) of warpage per panel was determined not by the breaking point of the glass (which in sheet form is a relatively flexible material) but rather by the tensile strength of the silicone seal that binds the two panes of glass together and to the curtain-wall frame. The digital model was then used to determine the correlating "unbent" dimensions for each flat panel that would translate to the correct shape after cold-warping on site, and these data were exported directly to the automated fabrication equipment that cut each piece of glass and aluminum to the required size. As is typical with prefabricated unit systems, each individual curtain-wall unit is framed on all sides by mullions of extruded aluminum, in this case about 75 millimeters wide by 200 deep (3 by 8 inches) to which the glass panels are structurally glazed with silicone sealant. The mullions are shaped in cross-section to interlock with the frames of adjacent units, providing additional structural stability and sealing the joint against water and air infiltration. Each unit is individually anchored to the building's concrete floor slab with adjustable aluminum brackets that can accommodate the allowable tolerances of the site-cast concrete frame.[7] To achieve cold-bending, workers partially installed each unit, anchoring it at the two bottom corners and one upper, and then physically pulled the fourth corner into place, connecting it to the slab to hold it in the final position. A sophisticated

7.17 InterActiveCorp Headquarters, New York City, USA. Gehry Partners, 2007.

laser-surveying system was used during installation to ensure that each unit was correctly located in the x, y, and z coordinates according to the digital model. Construction of the curtain wall was thus a combination of digital finesse and brute force.

The end result of this work is a building envelope with a visually indeterminate exterior character that seems to shift with environmental conditions: opaque under intense sun, translucent under overcast skies, nearly transparent at night when illuminated from within. The subtlety of the fading pattern on the fritted glass is more apparent from the interior, where one can experience up-close the parallactic effects of the gradient. The interior work spaces of the building are infused with a consistent, natural light, tempered by the partial shading provided by the ceramic frit pattern (and, if needed, translucent mesh shades can be lowered

from the ceiling to further reduce glare at work stations). From the outside the transitional zones of translucency between solid frit and transparency are often lost, and from a distance the façade reads, at times, simply as black stripes on a white background (or, as one critic put it, "a striated look recalling a Xerox machine that's running out of toner ink."[8]).

As the first building in New York City by the Pritzker-Prize-winning Gehry, the IAC Headquarters has received extensive media attention, garnering both positive and negative reviews. In a generally unfavorable assessment, *Newsday* critic Justin Davidson called the IAC Headquarters a "milky hulk on the Hudson," but offered at least faint praise for the building skin, writing: "The building reaches its apex of glamour in wretched weather. Fog and snow haze its edges and bleach its white skin

7.18 Office interior. InterActiveCorp Headquarters, New York City, USA. Gehry Partners, 2007.

7.19 Lobby interior, day. InterActiveCorp Headquarters, New York City, USA. Gehry Partners, 2007.

7.20 Lobby exterior, night. InterActiveCorp Headquarters, New York City, USA. Gehry Partners, 2007.

7.21 Johnson Wax Research Tower, Racine, Wisconsin, USA. Frank Lloyd Wright, 1950.

7.22 Johnson Wax Research Tower, left, 1950, and Administration Building, right, 1939, Racine, Wisconsin, USA. Frank Lloyd Wright.

whiter, so that it seems to be constantly evanescing and rematerializing."[9] Setting aside such interpretations of the building's visual and atmospheric effects, which are admittedly intriguing, it is perhaps most useful to view the IAC building as an example of the potential outcomes of deliberate and collaborative experimentation in the realms of materiality and curtain-wall fabrication.

* * *

Although not directly related, echoes of an earlier project—the Johnson Wax Research Tower by Frank Lloyd Wright—are evident in the IAC Headquarters. The projects are separated by nearly 60 years, but both are notable for their innovations in materiality and fabrication, particularly as these relate to building-envelope design. The two buildings, one in New York and the other in Racine, Wisconsin, are almost identical in height and share a pronounced horizontal banding of translucent glass on their façades. And although the two businesses couldn't be more different—IAC is a twenty-first-century conglomeration of dot-coms while Johnson Wax has its roots in the nineteenth century and oversees a vast empire of home-cleaning and pest-control products—the

companies' leaders shared an appreciation for architecture that compelled them to commission important buildings from the most celebrated architects of their times. In fact, Herbert F. Johnson, president of Johnson Wax and son of the company's founder, hired Frank Lloyd Wright on not one but three occasions: once for his own house (completed in 1937) and twice for Johnson Wax buildings (the original Administration Building, completed in 1939, and the Research Tower, added in1950).[10]

Among the much-noted innovations of Wright's two Johnson Wax buildings are the design of their tree-like structural systems and the use of hollow glass tubing to form translucent glazing. In the great central open space of the Administration Building, glass tubes are used mainly in double-layered skylights and clerestory windows to bring filtered light into the room from above.[11] The dominant feature here is, of course, the famous forest of tall, dendriform concrete columns that support the roof deck. Paul Goldberger calls it an "altogether spectacular room of light and swirling curves under a translucent ceiling."[12] However, to some degree, the glass assumes a background role, filling the leftover spaces between the broad, circular column caps. In contrast, at the adjacent Research Tower, glass tubes cover 70 percent of the

7.23 Johnson Wax Administration Building, Racine, Wisconsin, USA. Frank Lloyd Wright, 1939.

7.24 Johnson Wax Administration Building, Racine, Wisconsin, USA. Frank Lloyd Wright, 1939.

façade surface and constitute the primary expressive feature of the building, both inside and out, as well as the principal source of light. Wright described the Tower as "a sun-centered laboratory we now call the Helio-lab."[13] The following analysis will focus upon the distinctive building skin of the Research Tower, which expanded upon techniques first attempted in the Administration Building.

As a vertical counterpoint to the low, horizontal Administration Building, the Tower stands 47 meters tall (153 feet). Its structure consists of a central reinforced-concrete core, about 4 meters in diameter (13 feet), from which the concrete floor slabs cantilever. This core extends 16.5 meters (54 feet) into the ground to stabilize the Tower against wind-induced sway. The floor plates of the building's 14 levels alternate between squares with rounded corners reaching to the perimeter walls and circular mezzanines set back from the perimeter, creating double-height spaces that can

accommodate tall lab equipment and facilitate the penetration of natural light deeper into the labs. The concrete core contains mechanical shafts, an elevator, a staircase, and bathrooms, and serves as the sole vertical structure, meaning that the laboratory spaces are column-free and that the exterior skin is truly a curtain wall.

For the design of this curtain wall, Wright returned to the material palette he had used previously for the Administration Building: red brick and Pyrex glass tubes. Horizontal bands of bricks, edged with Kasota limestone, cover the edges of each square floor plate and extend vertically about 1.2 meters (4 feet). The remainder of each double-height wall consists of stacked glass tubes, each measuring 5 centimeters in diameter (2 inches). Pyrex, a non-discoloring glass with high thermal resistance, was first produced by Corning Glass Works in 1915 and had been

Laboratory interior. Johnson Wax Research Tower, Racine, Wisconsin, USA. Frank Lloyd Wright, 1950.

7.26   Johnson Wax Research Tower, left, 1950, and Administration Building, right, 1939, Racine, Wisconsin, USA. Frank Lloyd Wright.

previously used primarily for cookware and laboratory instruments.[14] Wright invented a new application for the material as architectural glazing, in cylindrical tube form, resulting in a striking visual effect but also, as we shall see, in significant performance issues. For the Research Tower façade, Corning supplied about 28 linear kilometers (17.5 miles) of Pyrex tubing. Wright also had to invent a method of installing and supporting the glass tubes and a technique for sealing the kilometers of joints between them. His solution was to use vertical, custom-formed aluminum mullions, spaced to individually support the ends of each tube while also creating a 6-millimeter wide (¼ inch) horizontal gap between each tube that could be filled with a flexible weatherproofing gasket to

seal the joint. Corning engineers designed the compressible gaskets, which were made of Koroseal (a polyvinyl chloride product), specially for the Tower. Installation of the tubes was extremely labor-intensive: each tube and gasket was placed individually by hand. The tubes form the outer layer of a double-skin envelope, intended as "temperature insurance" in Wright's words.[15] Separated by an air space of about 125 millimeters (5 inches), the inner layer consists of large panes of 6-millimeter thick (¼ inch) transparent plate glass that are hinged at the interior side of each aluminum mullion to facilitate cleaning.

The fact that such a novel system had never been built before led to some unusual situations on the construction site. A dispute

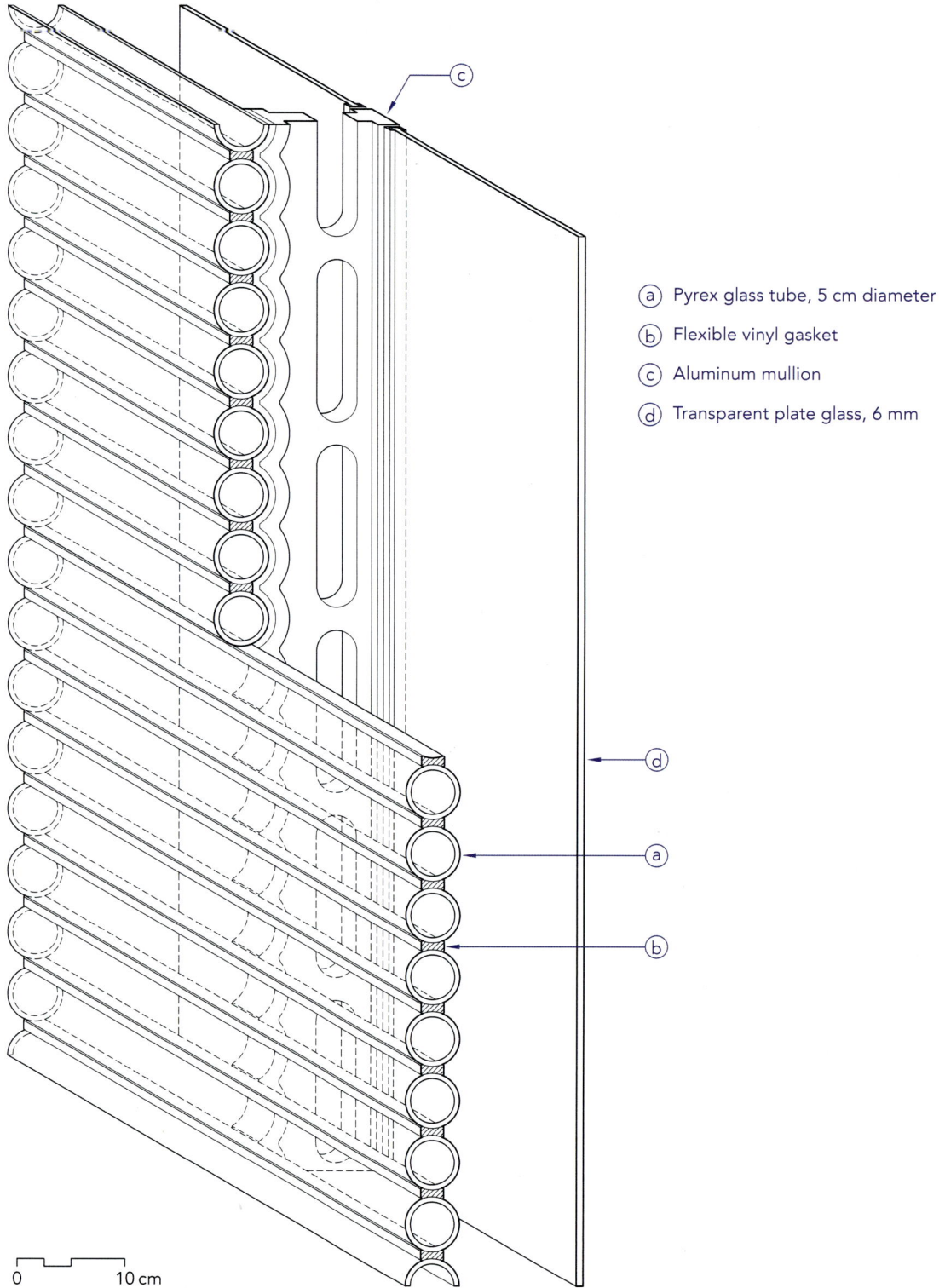

(a) Pyrex glass tube, 5 cm diameter

(b) Flexible vinyl gasket

(c) Aluminum mullion

(d) Transparent plate glass, 6 mm

0   10 cm

7.27 Glass wall assembly. Johnson Wax Research Tower.

7.28 Night view. Johnson Wax Research Tower, Racine, Wisconsin, USA. Frank Lloyd Wright, 1950.

7.29 Detail view, glass-tube wall. Johnson Wax Administration Building, Racine, Wisconsin, USA. Frank Lloyd Wright, 1939.

arose over which trade should have responsibility for installing the glass-tube system. Originally masons were given the job, based on the premise that stacking glass tubes was similar to laying bricks. Glaziers, who felt they should install all of the glass on the building, regardless of its form, called a strike in response. An agreement was eventually reached which stipulated that each trade would install the tubes on alternate floors, resulting in an unofficial race to completion that ultimately benefitted the construction schedule.[16]

Although the Pyrex tubes were manufactured with clear glass, their cylindrical curvature generates a refraction of light resulting in an ethereal, translucent quality. The tubes produce what Henry Plummer has called "astigmatic effects," or, in other words, the axially spreading of images due to light manipulation. Plummer further writes that "as glinting reflections and refractive distortions are superimposed onto views, glass attains a magical presence that is there one moment, and gone the next."[17] The tubes obscure direct vision through the curtain wall, but transmit ample diffuse light. Under certain lighting conditions, the Tower's façade takes on the appearance of a faded X-ray radiograph, revealing a ghostly image of the building's interior skeleton.[18] At night, the

Tower emits a soft glow that abstractly reveals a sense of the depth of space within.

The Johnson Wax Research Tower has been controversial in many respects, from its unconventional cantilevered structure to its unprecedented glass-tube curtain wall. This latter component became notorious for leaking, as the 30,000 linear meters (100,000 feet) of gaskets between glass tubes inevitably dried out and failed, and researchers who worked in the building have recalled keeping five-gallon buckets handy in their laboratories to collect rainwater.[19] In 1958, silicone sealant—a new product from Dow Corning used for the first time here—was successfully applied to seal the tube joints.[20] During the design phase, the company charged with designing the air-conditioning system for the building had complained that "since the construction of both the walls and the windows is new, there is no published data on the heat transfer factors and solar radiation effect through the glass."[21] Indeed, the building envelope, even with its double-skin configuration, proved to be thermally inadequate to maintain comfortable temperatures in the labs when faced with the seasonal temperature swings of the Wisconsin climate. Comically, at one point chemists were issued sunglasses to deal with intense

7.30 Johnson Wax Research Tower, Racine, Wisconsin, USA. Frank Lloyd Wright, 1950.

7.31 Johnson Wax Research Tower, left, 1950, and Administration Building, right, 1939, Racine, Wisconsin, USA. Frank Lloyd Wright.

glare until special translucent curtains could be installed in the labs.[22] More serious issues, however, eventually led Johnson Wax to stop using the Research Tower altogether. The building's useful life as a research facility ended just 31 years after its opening, when Johnson Wax acquired an existing building nearby and moved its research operations there. This decision was due in part to the inherent difficulties in expanding or upgrading the laboratory spaces in the slender Tower and also to the building's inadequate circulation system. With only a single 75-centimeter wide (30 inch) staircase in its core, the Tower did not meet current building code requirements for emergency egress in case of fire. The possible insertion of a second staircase, which would satisfy the code, was deemed unfeasible given the already tight core configuration, and any external addition was considered an unacceptable alteration of Wright's original design. The Tower has therefore been unoccupied since 1981.

Despite its problems and eventual closure, the Research Tower remains an important part of the Johnson Wax complex (the tower is still illuminated for several hours each night) and looms large as an icon of mid-twentieth-century modernism. One scientist who worked in the building for 30 years acknowledged its shortcomings but felt that Wright's architecture contributed something important and intangible to the company's success during that time, remarking that "because [the Tower] was unique, it made everybody think a little different."[23] In 1974, both the Administration Building and the Research Tower were named in the US National Register of Historic Places, and SC Johnson (as the company is now known) has confirmed its commitment to the Tower, even in its current state of disuse.[24] Fisk Johnson, the current CEO of the company, has said, "The Tower stands as a strong symbol of SC Johnson ingenuity . . . We will maintain the Tower and light it every night as a beacon to our community."[25] Although no longer a functional research facility, the luminous Tower with its never-duplicated glass skin has taken on a second life as a unique kind of monument, falling into its own category somewhere between public art and architecture.

Acknowledgements

It has been a pleasure working on this book with the editorial team at Routledge. Special thanks go to Francesca Ford, Laura Williamson, and Alanna Donaldson for their professionalism and enthusiasm for this project.

The research for this book has been generously supported by grants from the Campus Research Board at the University of Illinois at Urbana-Champaign, a Creative Research Award from the College of Fine & Applied Arts at the University of Illinois, and travel grants from the William and Flora Hewlett Foundation and from the Alan and Leonarda Laing Foundation. In addition to these resources, I am grateful for access to one of the world's great library systems at the University of Illinois. The breadth and depth of its collections, as well as the expertise and helpfulness of its staff, have been invaluable in the progression of my research over the years.

This book is dedicated to family and friends whose encouragement has aided my professional and academic development in countless ways and without whom this work would not be possible. Thank you.

Scott Murray
Urbana, Illinois

Summary of Primary Architects

(Listed chronologically by project)

Steiff Factory Building, 1903
Richard Steiff (1877–1939), Giengen, Germany

Maison de Verre, 1932
Pierre Chareau (1883–1950) and Bernard Bijvoet (1889–1979), Paris, France

Steuben Glass Building, 1937
William Platt (1897–1984) and Geoffrey Platt (1905–1985), New York City, USA

Museum of Modern Art, 1939
Philip Goodwin (1885–1958) and Edward Durell Stone (1902–1978), New York City, USA

Johnson Wax Research Tower, 1950, and Beth Sholom Synagogue, 1959
Frank Lloyd Wright (1867–1959), Spring Green, Wisconsin, and Scottsdale, Arizona, USA

Beinecke Rare Book Library, 1963
Gordon Bunshaft (1909–1990), Skidmore, Owings & Merrill, New York City, USA

Kunsthaus Bregenz, 1997
Peter Zumthor (1943–), Haldenstein, Switzerland

Kursaal Congress Center, 1999
Rafael Moneo (1937–), Madrid, Spain

Maison Hermès, 2001
Renzo Piano (1937–), Renzo Piano Building Workshop, Genoa, Italy, and Paris, France
www.rpbw.com

Laban Dance Center, 2003
Jacques Herzog (1950–) and Pierre de Meuron (1950–), Herzog & de Meuron, Basel, Switzerland
www.herzogdemeuron.com

LVMH Building, 2004
Kengo Kuma (1954–), Kengo Kuma and Associates, Tokyo, Japan
www.kkaa.co.jp

Higgins Hall Insertion, 2005
Steven Holl (1947–), Steven Holl Architects, New York City, USA
www.stevenholl.com

InterActiveCorp Headquarters, 2007
Frank Gehry (1929–), Gehry Partners LLP, Los Angeles, USA
www.foga.com

Notes

PREFACE

i   Joan Ockman, "Light Construction Symposium" (1995) in Todd
    Gannon, ed., *The Light Construction Reader* (New York: Monacelli
    Press, 2002), 66.

INTRODUCTION

1   Hal Foster, "The ABCs of Contemporary Design," *October* 100 (Spring
    2002), 194.

2   Even Blau, "Tensions in Transparency," *Harvard Design Magazine* 29
    (Fall/Winter 2008–9), 29.

3   Anthony Vidler, *The Architectural Uncanny* (Cambridge, Mass.: MIT
    Press, 1992), 217.

4   Steven Holl, "Nelson-Atkins Museum of Art," in Michael Bell and Jeannie
    Kim, ed., *Engineered Transparency: The Technical, Visual, and Spatial
    Effects of Glass* (New York: Princeton Architectural Press, 2009), 97.

5   An early example is the Maison de Verre of 1932, a project analyzed in
    detail in Chapter 1 of this book. Although the Maison de Verre was not
    initially widely appreciated, it has now attained iconic status. The critic
    Nicolai Ouroussoff notes that this building "had a profound impact on
    generations of architects who were seeking to free themselves from the
    rigid orthodoxies of mainstream Modernism." Nicolai Ouroussoff, "The
    Best House in Paris," *New York Times* (August 26, 2007), Arts & Leisure,
    22.

6   Walter Gropius, *The New Architecture and the Bauhaus* (Cambridge,
    Mass.: MIT Press, 1965), 29.

7   Arthur Korn, *Glass in Modern Architecture* (London: Barrie & Rockcliff,
    1968), 6. First published in German as *Glas im Bau und als
    Gebrauchsgegenstand* in 1929.

8   Ludwig Mies van der Rohe, *Fruhlicht* (1922), as translated in Peter
    Carter, *Mies van der Rohe at Work* (London: Phaidon Press, 1999), 18.

9   Colin Rowe and Robert Slutzky, "Transparency: Literal and
    Phenomenal," *Perspecta* 8 (1963): 45–54. A sequel was published eight
    years later: Rowe and Slutzky, "Transparency: Literal and Phenomenal
    . . . Part 2," *Perspecta* 13/14 (1971): 287–301.

10 Detlef Mertins, *Modernity Unbound: Other Histories of Architectural
    Modernity* (London: AA Publications, 2011), 11.

11 See the essay "Transparency: Autonomy and Relationality" in Mertins,
    Modernity Unbound, 70–87. Originally published in *AA Files* 32 (1997),
    3–11.

12 Two books by Henry Plummer provide valuable insights into the
    historical significance of light in architecture: *Masters of Light* (Tokyo:
    A+U Publishing Co., 2003) and *The Architecture of Natural Light* (New
    York: Monacelli Press, 2009).

13 Another intriguing example of a translucent building envelope that
    appears to fully anticipate the modern aesthetic is the 1891 double-skin
    glass house built by William van der Heyden in Yokohama, Japan. See
    Pedro Guedes, "Presciently Translucent in Yokohama, 1891," *ARQ:
    Architectural Research Quarterly* 9 (2005), 69–79.

14 Le Corbusier, *Towards a New Architecture* (New York: Dover, 1986), 37.
    First published in French as *Vers un Architecture* in 1923.

15 Louis Kahn, "Space and the Inspirations" (1967), in Robert Twombly,
    ed., *Louis Kahn: Essential Texts* (New York: WW Norton, 2003), 225.

16 Steven Holl, *Questions of Perception: Phenomenology of Architecture*
    (San Francisco: William Stout, 2006), 83. First published by A+U
    Publishing Co., 1994.

17 Holl, "Nelson-Atkins Museum of Art," in *Engineered Transparency*, 97.

18 Richard P. Feynman, *QED: The Strange Theory of Light and Matter*
    (Princeton, NJ: Princeton University Press, 2006), 16. First published in
    1985.

19 Juhani Pallasmaa, *The Eyes of the Skin: Architecture and the Senses*
    (London: Academy Editions, 1996), 46.

1 SOLIDIFIED LIGHT

1   Steven Holl, "Nelson-Atkins Museum of Art," in Michael Bell and
    Jeannie Kim, ed., *Engineered Transparency: The Technical, Visual, and
    Spatial Effects of Glass* (New York: Princeton Architectural Press, 2009),
    97.

2   See Kenneth Frampton, *Steven Holl Architect* (Milan: Electa
    Architecture, 2003) and Steven Holl, *Urbanisms: Working with Doubt*
    (New York: Princeton Architectural Press, 2009).

3   Holl's ability to secure a string of increasingly significant commissions
    in the early 2000s must have only been helped by *Time* magazine
    naming him "America's Best Architect" in July 2001.

4 The Dalsaces were repeat clients: the couple had also previously given Chareau his first private commission, in 1918, to decorate their two-room apartment. See Marc Vellay and Kenneth Frampton, *Pierre Chareau: Architect and Craftsman* (London: Thames and Hudson, 1985), 34. The Maison de Verre stayed in the Dalsace family, used to varying degrees over the years as a domicile, until it was sold in 2006 to Robert Rubin, an entrepreneur, scholar, and Director of the Society of Architectural Historians, who then embarked on a careful restoration of the building, returning several of its features that had been altered over time to their original configurations. See Nicolai Ouroussoff, "The Best House in Paris," *New York Times* (August 26, 2007), Arts & Leisure, 22–3.

5 "The Best House in Paris," 1. Kenneth Frampton also addresses this phenomenon, writing that the Maison de Verre "became at once part of an underground tradition; its immediate influence limited to a select few who were sympathetic to its creation," including Le Corbusier. See Kenneth Frampton, "Maison de Verre," *Perspecta* 12 (1969), 80.

6 "Une Maison de Verre," *Glace et Verre* 17 (1930), 19–20, as quoted in Marc Vellay, *La Maison de Verre* (Tokyo: ADA Edita, 1988), 10.

7 As quoted in *Pierre Chareau: Architect and Craftsman*, 239. Dalsace goes on to praise Chareau's apparently empathetic connection to his patron: "The whole house has been created under the influence of friendship, in complete affectionate understanding." A satisfied client, indeed.

8 *Pierre Chareau: Architect and Craftsman*, 242–3.

9 Frampton, "Maison de Verre," 77.

10 Component dimensions are noted in *La Maison de Verre*, 11–12.

11 See Frampton, "Maison de Verre," 77. Frampton notes, "It is a measure of Chareau's clients' courage that they were willing to adopt such an unproven material for the enclosure of their house."

12 John Gloag, *The Place of Glass in Buildings* (London: George Allen & Unwin, 1943), 14.

13 Historians' assessments of the Maison de Verre have been previously noted. Among other honors, the Higgins Hall Insertion won the 2007 Architecture Honor Award from the New York Chapter of the American Institute of Architecture. And Kenneth Frampton, who has written extensively of the Maison de Verre, praised the "economically heroic character" of Holl's Pratt building, calling it "an ingenious piece of urban infill . . . that is equal in terms of its significant distribution of space to any comparable school of architecture in the U.S." See Kenneth Frampton, "Prometheus Bound and Unbound," *Domus* 896 (October 2006).

14 See Fred A. Bernstein, "Where Old and New Collide," *Metropolis* (February 2006), 114–17.

15 Ibid., 114.

16 See Steven Holl, *Urbanisms: Working With Doubt* (New York: Princeton Architectural Press, 2009), 52–57.

17 R.A. Heintges & Associates, for whom the author worked from 2001 to 2005, is a curtain-wall consulting firm established in New York City in 1989 by Robert A. Heintges. The firm also collaborated with Steven Holl Architects on the channel-glass curtain wall for the Nelson-Atkins Museum of Art in Kansas City, Missouri. See www.heintges.com. W&W Glass Systems has been involved in many high-profile custom glazing projects in New York City, including 40 Bond Street by Herzog & de Meuron and 40 Mercer Street by Jean Nouvel. See www.wwglass.com.

18 From a Firm Profile provided by Steven Holl Architects (2011).

2 ART HOUSE CINEMA

1 Zumthor won the design competition for the project in 1990. Construction began in 1994; the Kunsthaus opened on July 25, 1997, and won the Mies Award the following year. From http://www.miesarch.com: "The Prize is a joint initiative of the European Commission and the Fundació Mies van der Rohe. The principal objectives are to recognize and commend excellence in the field of architecture and to draw attention to the important contribution of European professionals in the development of new concepts and technologies."

2 "Mystical Presence," *Architectural Review* (December 1997), 46. While reaction in the architectural press was generally positive, there have been reviews critical of the building's austerity. For instance, Victoria Newhouse writes: "Zumthor's Kunsthaus is suitable only for contemporary art capable of animating its severe setting. In this it may prove to be as limited in what it can successfully exhibit as Peter Eisenman's Wexner Center." Victoria Newhouse, *Towards a New Museum* (New York: Monacelli Press, 2006), 72.

3 Kunsthaus Bregenz is also notable for its innovative, low-energy thermal

conditioning system. See Kiel Moe, *Thermally Active Surfaces in Architecture* (New York: Princeton Architectural Press, 2010), 132–145.

4 Paul Scheerbart and Bruno Taut, *Glass Architecture and Alpine Architecture*, ed. Dennis Sharp, trans. James Palmes and Shirley Palmer (New York: Praeger, 1972), 51. This volume contains English translations of Scheerbart's *Glasarchitektur* (1914) and Taut's *Alpine Architektur* (1919).

5 "Mystical Presence," 46.

6 Richard Ingersoll, "Light Boxes," *Architecture* (October 1997), 99.

7 Peter Zumthor, *Kunsthaus Bregenz* (Ostfildern: Hatje, 1999), 13.

8 Ibid., 13.

9 Diminutive, perhaps, relative to two other well-known museums which also opened in 1997: Richard Meier's sprawling 360,000-square foot J. Paul Getty Center in Los Angeles and Frank Gehry's 256,000-square foot Guggenheim Museum in Bilbao, Spain.

10 From an interview by the author with employees at Fällander Glass, Fällanden, Switzerland, June 12, 2006.

11 A comprehensive list of past exhibitions may be viewed at the Kunsthaus Bregenz website, http://www.kunsthaus-bregenz.at.

12 For more on Holl's Nelson-Atkins Museum project, see Scott Murray, *Contemporary Curtain Wall Architecture* (New York: Princeton Architectural Press, 2009), 198–205.

13 Glenn Lowry, *The New Museum of Modern Art* (New York: Museum of Modern Art, 2005), 10.

14 For more on the founding of MoMA, see Harriet Bee and Michelle Elligott, ed., *Art In Our Time: A Chronicle of the Museum of Modern Art* (New York: Museum of Modern Art, 2004).

15 This similarity is noted by Victoria Newhouse in *Towards a New Museum*, 149, where she writes: "This and several other of Howe and Lescaze's schemes were more innovative than anything MoMA has ever built." For a more detailed description of the Howe & Lescaze schemes, see Robert A.M. Stern, Gregory Gilmartin, and Thomas Mellins, *New York 1930: Architecture and Urbanism Between the Two World Wars* (New York: Rizzoli, 1987), 141–144.

16 Hitchcock and Johnson's 1932 International Style exhibition became MoMA's first traveling exhibition. Hitchcock and Johnson also wrote the accompanying book, *The International Style*.

17 The interesting saga of MoMA's architect selection process is discussed in detail in Dominic Ricciotti, "The 1939 Building of the Museum of Modern Art: The Goodwin–Stone Collaboration," *American Art Journal* 66, No. 3 (Summer 1985), 50–76.

18 *Art In Our Time*, 54. Others disagreed, including Philip Johnson, who later called the building "a watered [down] version of an International Style building," as quoted in *Towards a New Museum*, 151.

19 Libbey-Owens-Ford Catalog, *Sweet's Catalog File*, 1941.

20 See "The 1939 Building of the Museum of Modern Art: The Goodwin–Stone Collaboration," 68–69.

21 The steel-sash strip windows at the fourth and fifth floors were also replaced with aluminum windows and new glass.

22 Lewis Mumford, "The Skyline: Growing Pains—The New Museum," *The New Yorker* (June 3, 1939).

23 The author was part of the design team for this restoration, serving as a project manager in the office of R.A. Heintges.

24 "Museum of Modern Art," *Architectural Forum* (August 1939), 116.

25 Talbot Hamlin, "Modern Display for Works of Art," *Pencil Points* (September 1939), 615.

3 CRYSTALLIZATION

1 Acknowledging the unusual aqueous appearance of the glass building and the fact that most of the unmarried young women in Giengen were employed at Steiff, locals apparently developed a nickname for the building: the "virgin aquarium." See Henning Meyer, "Historischer Fabrikbau der Firma Steiff in Giengen," *Baumeister* (November 2003), 90–95.

2 See Anke Fissabre and Bernhard Niethammer, "The Invention of the Glazed Curtain Wall in 1903 – The Steiff Toy Factory," *Proceedings of the Third International Congress on Construction History*, Cottbus (May 2009), 595–602.

3 See Georg Kohlmaier and Barna von Satory, *Houses of Glass: A Nineteenth-Century Building Type*, trans. John C. Harvey (Cambridge, Mass.: MIT Press, 1986). First published in German in 1981.

4 Each of the three buildings used glass curtain wall as the primary enclosure, resulting in a similar appearance from the exterior. The later buildings, however, used timber structural systems instead of steel.

5 For more on the technologies of the curtain wall and its historical

development, see Scott Murray, *Contemporary Curtain Wall Architecture* (New York: Princeton Architectural Press, 2009).

6 See "Historischer Fabrikbau der Firma Steiff in Giengen." Eisenwerk München, likely collaborators in the overall design with Richard Steiff, produced the official structural shop drawings that were submitted for approval to the local authorities.

7 Insulating glass (consisting of prefabricated units with a sealed air cavity between two or more sheets of glass) had not yet been invented. The double-skin configuration in some ways prefigures insulating glass, as it was based on the same principles, but included a deeper cavity and required on-site assembly.

8 The Steiff Building is largely absent from the canonic histories of modernism, such as Henry-Russell Hitchcock's *Modern Architecture: Romanticism and Reintegration*, Reyner Banham's *Theory and Design in the First Machine Age*, and Kenneth Frampton's *Modern Architecture: A Critical History*.

9 The temple analogy is also noted in Alan Colquhoun, *Modern Architecture* (Oxford: Oxford University Press, 2002), 66.

10 Jacques Herzog & Pierre de Meuron received the commission by winning an international design competition in 1997. Laban is their second project in the UK, following their 2000 conversion of the Bankside Power Station into the new Tate Modern Gallery in London, also the result of a design competition. In 2005, the architects were again commissioned by Tate Modern, this time to design another major expansion that is scheduled for completion in 2016.

11 The Stirling Prize is presented annually "to the architects of the building that has made the greatest contribution to the evolution of architecture in the past year." To be eligible, a project must have been either built or designed in the UK. See http://www.architecture.com/Awards/RIBAStirlingPrize.

12 The curvature is said to be a reference to the nearby St. Paul's Church designed by Thomas Archer and completed in 1730. The Laban forecourt was designed in collaboration with Vogt Landscape Architects. See Luis Fernández-Galiano, ed. *Herzog & de Meuron 1978–2007* (Madrid: Arquitectura Viva, 2007), 170.

13 Rafael Moneo, *Theoretical Anxiety and Design Strategies in the Work of Eight Contemporary Architects* (Cambridge, Mass.: MIT Press, 2004), 363–364.

14 Giles Reid, "Dancing Architecture," *Domus* (April 2003), 67.

4 COMPOUND LENS

1 Jury Citation for Renzo Piano, 1998 Laureate, Pritzker Architecture Prize, http://www.pritzkerprize.com/laureates/1998/jury.html.

2 For the complete works, see Philip Jodido, *Piano: Renzo Piano Building Workshop, 1966 to Today* (Cologne: Taschen, 2008).

3 RPBW designed an addition to Maison Hermès, completed in 2006, which expanded the length of the building but was designed to appear seamless from the exterior.

4 This is a strategy similar to that of Kengo Kuma's LVMH Building in Osaka (see Chapter 5), though resulting in an entirely different aesthetic.

5 Quoted from a project statement provided by the architect.

6 Examples of this global trend are too numerous to list comprehensively, but, in addition to the buildings named in this chapter, they include Christian de Portzamparc's LVMH Tower (1999) in New York, OMA's Prada Epicenter Stores (2001 and 2004) in New York and Beverly Hills, Herzog & de Meuron's Prada Flagship Store (2003) in Tokyo, as well as Toyo Ito's Tod's Omotesando Building (2004) and Mikimoto Ginza (2005) in Tokyo.

7 Critic Raul Barreneche is one who has compared Maison Hermès to Maison de Verre, calling Piano's cladding design "a riff on history" that is less inventive than Piano's previous work. See Raul Barreneche, "Fashion House Hermès' Flagship Store by Renzo Piano," *Architectural Record*, October 2002, 166–170.

8 Interestingly, the Platts were a family of architects: William and Geoffrey Platt were the sons of architect and landscape designer Charles Adams Platt, and William Platt had a son, named Charles Platt, who also became an architect and was founder of the New York firm Platt Byard Dovell White.

9 Lewis Mumford, "The Sky Line: Gardens and Glass," *The New Yorker*, October 16, 1937, 68–70. Mumford's review was generally positive, with his main criticism aimed at the few ornamental stone flourishes on the façade, which he described as "piously attaching lumps of sculpture to a building in which they have no integral part."

10 As quoted in "The Week in Science: First Glass House," *New York Times*, September 27, 1936. Platt seems to be echoing Walter Gropius who wrote of the "growing preponderance of voids over solids" and the concept of walls as "mere screens stretched between the upright

columns." See Walter Gropius, *The New Architecture and the Bauhaus* (Cambridge, Mass.: MIT Press, 1965), 26–29.

11 For more on Harrison's Corning Building and other midcentury towers, see Scott Murray, *Contemporary Curtain Wall Architecture* (New York: Princeton Architectural Press, 2009).

12 The history of the Steuben Building site, dating back to 1872, is carefully delineated by Christopher Gray in "An Architectural Barometer," *New York Times*, September 1, 2011. Gray also points out that in 1960 "the Fifth Avenue Association gave [Winston] an award for the best altered building of the year" and that "another winner was the modernist tower across the street—the new Corning Glass Building."

13 Ada Louise Huxtable, "Some New Faces on Fifth Avenue," *New York Times*, November 8, 1964.

14 "The Week in Science: First Glass House," *New York Times*, September 27, 1936.

15 Robert A.M. Stern, Gregory Gilmatin, and Thomas Mellins, *New York 1930: Architecture and Urbanism Between the Two World Wars* (New York: Rizzoli, 1987), 378.

16 The design was also published in *The Inland News and Architect Record* in January 1898.

17 Detlef Mertins, "Bioconstructivisms," in Michael Bell and Jeannie Kim, ed., *Engineered Transparency—The Technical, Visual, and Spatial Effects of Glass* (New York: Princeton Architectural Press, 2009), 34.

18 For a detailed history of Luxfer, see Dietrich Neumann, "'The Century's Triumph in Lighting': The Luxfer Prism Companies and their Contribution to Early Modern Architecture," *Journal of the Society of Architectural Historians* 54(1), March 1995, 24–53.

5 GEOLOGY

1 As quoted in Carole Herselle Krinsky, *Gordon Bunshaft of Skidmore, Owings & Merrill* (Cambridge, Mass.: MIT Press, 1988), 146.

2 Bunshaft first joined the New York office of Skidmore and Owings in 1937. He left the firm to serve in the US military from 1942 to 1946 and then returned to join the office that had by then become Skidmore, Owings & Merrill. Bunshaft was with SOM until 1979. He won the Pritzker Architecture Prize in 1988. For further information, see Detlef

Mertins, ed. "Gordon Bunshaft interviewed by Betty J. Blum," *SOM Journal* 3 (2004).

3 In an overview of Bunshaft's career, the critic Paul Goldberger wrote that his oeuvre is one that "ranks as one of the major architectural achievements of the [twentieth] century." Paul Goldberger, "Gordon Bunshaft: A Man Who Died Before His Time?" *New York Times*, August 19, 1990.

4 For example, the Bunshaft-designed Manufacturers Hanover Trust Company, a bank headquarters in New York City, is a classic modernist example of transparency deployed as a guiding principle. The building, which was named a New York City Landmark in 1997, is famous for the clear views of the bank's interior spaces—and notably its vault—offered to passersby on Fifth Avenue through a monumental transparent glass curtain wall.

5 Antony Hobson, *Great Libraries* (New York: Putnam's Sons, 1970), 232. Carole Herselle Krinsky cites Hobson's praise as well, calling it part of "Bunshaft's favorite description of the library." Krinsky also notes strong criticism of the building by Vincent Scully and others. See *Gordon Bunshaft of Skidmore, Owings & Merrill*, 143–145.

6 See Patrick Loughran, *Failed Stone: Problems and Solutions with Concrete and Masonry* (Basel: Birkhäuser, 2007), 10–21.

7 Henry Plummer describes the Beinecke's interior as "a half-lit, dream-like space with its literary treasures" and "its precious objects bathed in mystical light." Henry Plummer, *Masters of Light* (Tokyo: A+U Publishing Co., 2003), 162.

8 For further information, see David Dernie, *New Stone Architecture* (London: Laurence King, 2003), 21–22.

9 Rafael Moneo, *Remarks on 21 Works* (New York: Monacelli Press, 2010), 508–509.

10 From a project statement provided by Kengo Kuma and Associates.

11 Arthur Korn, *Glass in Modern Architecture* (London: Barrie & Rockcliff, 1968), 6. First published in German as *Glas im Bau und als Gebrauchsgegenstand* in 1929.

12 From Kengo Kuma's Introduction in Botond Bognar, *Material Immaterial: The New Work of Kengo Kuma* (New York: Princeton Architectural Press, 2009), 10–11.

13 More information on these projects and other recent works can be found in *Material Immaterial: The New Work of Kengo Kuma*.

14 See Francisco Asensio Cerver, *The Architecture of Glass: Shaping Light* (New York: Hearts Books International, 1997), 158–169.

15 Josep Lluís Mateo, "New Headquarters for the Deutsche Bundesbank," *A+U: Architecture and Urbanism* 447, Dec. 2007, 24–31.

6 BIOLUMINESCENCE

1 Moneo was selected as architect in 1990 from a shortlist of six internationally renowned architects by the municipality of San Sebastián. The new building occupies the site of the former Gran Kursaal (from which it takes its name), a casino that was built in 1921 and demolished in 1972.

2 Rafael Moneo, *Remarks on 21 Works* (New York: Monacelli Press, 2010), 377.

3 Rafael Moneo, *The Freedom of the Architect* (Ann Arbor: University of Michigan Press, 2002), 30.

4 The specific products used were: PPG Starphire Glass (low-iron), Dupont Butacite (PVB), and AFG/Glaverbel Flutex (textured glass).

5 From an interview by the author with representatives of Cricursa in Barcelona, Spain, June 16, 2006.

6 As quoted in William J.R. Curtis, "A Conversation with Rafael Moneo," *El Croquis* 98 (2000), 9.

7 The team of Rogers Marvel and Ken Smith won a 2002 design competition for the new plaza, which replaced an existing underutilized plaza that originally opened on the site in 1972.

8 See David W. Dunlap, "An Elevated Plaza Finally Worth Going Up To See," *New York Times* (October 19, 2005).

9 From the Machado and Silvetti website: http://www.machado-silvetti. com/projects/dewey_square/index.php.

10 Construction of the Beth Sholom Synagogue was overseen by Haskell Culwell, the general contractor who had previously built three other Wright projects, including the Price Tower (1952) in Bartlesville, Oklahoma. Wright recommended Culwell, who was based in Oklahoma, after four local contractors declined to bid on the Beth Sholom project, citing its unusual design. See Patricia Talbot Davis, *Together They Built a Mountain* (Lititz, Penn.: Sutter House, 1974), 78–87.

11 This height exceeded the town's established height limit of 20 meters (65 feet) and therefore required that an exception be granted by the local Zoning Board. See *Together They Built a Mountain*, 59.

12 Julia M. Klein calls Cohen "a visionary with chutzpah" and Beth Sholom "one of Wright's greatest architectural achievements." Klein, "The Rabbi and Frank Lloyd Wright," *Wall Street Journal* (December 22, 2009). During construction, Cohen was a fixture on the job site and often reported back to Wright his detailed observations and concerns. In one letter to Wright, for example, Cohen reports that the plastic panels being installed on the ceiling were misaligned. He writes, "The vertical joints . . . do not line up. They are staggered, but do not form a pattern, and consequently make a chaotic and bad impression . . . We are of the opinion that, unless this is corrected, we had better stop work on the building." Mortimer J. Cohen to Frank Lloyd Wright, 25 November 1958. Getty Research Institute.

13 A fascinating history of Beth Sholom's design and construction, including the personal relationships among its architects, owners, and builders, can be found in Patricia Talbot Davis's *Together They Built a Mountain*.

14 "Beth Sholom Synagogue: An American Synagogue" (pamphlet), (Elkins Park, Penn.: Beth Sholom Congregation, January 2008), 3.

15 Frank Lloyd Wright, *Modern Architecture: Being the Kahn Lectures of 1930* (Princeton, New Jersey: Princeton University Press, 1931), 38–9.

16 Wired glass, produced by pressing a steel wire mesh between two ribbons of semi-molten glass, was at the time considered an appropriate material for overhead glazing, as the mesh would help retain any fragments of glass in the event of breakage (though today's building codes would require laminated glass).

17 The blue plastic is noted in a 1954 letter from Wright to Cohen transmitting the original design sketches, as quoted in *Together They Built a Mountain*, 45. Also noted in "The Record Reports," *Architectural Record* (July 1954), 20.

18 See Peter Arnell and Ted Bickford, ed., *James Stirling, Buildings and Projects* (London: Architectural Press, 1984), 97–105.

19 See Jenna M. McKnight, "Prayer Pavilion of Light, Phoenix, Arizona," *Architectural Record* (June 2010), 169–73.

20 Brendan Gill, *Many Masks: A Life of Frank Lloyd Wright* (New York: Ballantine Books, 1988), 461.

7 FADE TO BLACK

1 IAC was followed in 2011 by Gehry's second New York building, the 76-story residential Beekman Tower (his tallest building to date).

2 As quoted in Paul Goldberger, "diller@gehry.nyc," *Vanity Fair* (June 2007).

3 Gehry and his client avoided what might have been the obvious solution: a dynamic media wall integrated into the façade, similar to that of another recent Manhattan office building, 745 Seventh Avenue (2001) by KPF Architects. IAC does have a state-of-the-art media wall, but it is located inside the lobby.

4 Shortly after opening in 2011, developers began marketing the Beekman Tower under a new name: "New York by Gehry."

5 See John Hockenberry, "Diller, Gehry, and the Glass Schooner on 18th Street," *Metropolis* (June 2007), 124–5.

6 See Reinhold Martin, "The Crystal World: Frank Gehry's IAC," *Harvard Design Magazine* (Fall 2007/Winter 2008), 4–13, and Paul Goldberger, "diller@gehry.nyc," *Vanity Fair* (June 2007).

7 The concrete structure was allowed to vary by up to 25 millimeters (1 inch) in any direction, an industry-standard tolerance, while the curtain-wall placement had to be accurate to within 3 millimeters (1/8 inch). See "IAC/InterActiveCorp Headquarters, New York," *Metals in Construction* (Spring 2007), 8–13.

8 Julie V. Iovine, "He'll Take Manhattan," *The Architect's Newspaper* (April 4, 2007).

9 Justin Davidson, "Architecture: The Glass Menagerie," *Newsday* (April 14, 2007).

10 A fourth project was discussed: at one point Johnson was interested in building 200 to 400 new houses for his company's employees, which he wanted Wright to design. The project never materialized. See Jonathan Lipman, *Frank Lloyd Wright and the Johnson Wax Buildings* (Minneola, New York: Dover, 1986), 143. However, the SC Johnson Company has recently extended its architectural patronage, hiring Foster + Partners to design a new public exhibition space, called Fortaleza Hall, which opened in 2010 adjacent to Wright's buildings. Foster's new structure shares a vocabulary of curvilinear form with its famous neighbors, but, with its transparent glass walls, stands in material contrast with Wright's translucent buildings.

11 The glass-tube skylights were particularly prone to leaking, and beginning in 1957 portions of the glass tubing were removed and replaced with sheets of fiberglass, corrugated to simulate the appearance of the tubes. See Lipman, *Frank Lloyd Wright and the Johnson Wax Buildings*, 169. For a personal account of the challenging con-struction of the glass-tube skylights by one of Wright's apprentices, see Edgar Tafel, *Years With Frank Lloyd Wright: Apprentice to Genius* (New York: Dover, 1979), 174–89.

12 Paul Goldberger, *Why Architecture Matters* (New Haven, Conn.: Yale University Press, 2009), 11.

13 Frank Lloyd Wright, "Helio-Laboratory for Johnson Wax Co.," *Architectural Forum* (January 1951), 77.

14 During the design of the Administration Building, Wright had considered other types of glass, including glass blocks and cast-glass panels, but was convinced to work with Corning when it showed a willingness to collaborate with Wright and to engage in the necessary experimentation with its Pyrex tubes. See Lipman, *Frank Lloyd Wright and the Johnson Wax Buildings*, 65.

15 Wright, "Helio-Laboratory for Johnson Wax Co.," 77.

16 Ibid., 156.

17 Henry Plummer, *The Architecture of Natural Light* (New York: Monacelli Press, 2009), 82–3.

18 Beatriz Colomina writes about the metaphor of the X-ray as it pertains to the work of Mies and other modern architects: "The X-ray logic, as it has been absorbed by modern architecture, culminates in a dense cloud of ghostly shapes." Beatriz Colomina, "Unclear Vision: Architectures of Surveillance," in Michael Bell and Jeannie Kim, ed., *Engineered Transparency: The Technical, Visual, and Spatial Effects of Glass* (New York: Princeton Architectural Press, 2009), 89.

19 Mark Hertzberg, *Frank Lloyd Wright's SC Johnson Research Tower* (San Francisco: Pomegranate, 2010), 59.

20 See Lipman, *Frank Lloyd Wright and the Johnson Wax Buildings*, 164.

21 From a 1945 memorandum by the Carrier Corporation, as quoted in Hertzberg, *Frank Lloyd Wright's SC Johnson Research Tower*, 58.

22 Ibid., 58.

23 Dr. Don Whyte, as quoted in Hertzberg, *Frank Lloyd Wright's SC Johnson Research Tower*, 55.

24 The original Johnson Wax Administration Building remains the company's international headquarters and receives about 4,500 visitors per year who tour the complex and learn about its history. Due to the code issues described above, the tour does not include a visit to the Research Tower.

25 As quoted in Hertzberg, *Frank Lloyd Wright's SC Johnson Research Tower*, 71.

Bibliography

INTRODUCTION: TRANSPARENCY / TRANSLUCENCY / OPACITY

Bell, Michael and Jeannie Kim, ed. *Engineered Transparency: The Technical, Visual, and Spatial Effects of Glass*. New York: Princeton Architectural Press, 2009.

Blau, Eve. "Tensions in Transparency: Between Information and Experience." *Harvard Design Magazine* 29, Fall/Winter 2008–9, 29–37.

Davidson, Cynthia. "Reflections on Transparency: An Interview with Terence Riley." *Any* 9, 1994, 56–57.

Feynman, Richard P. *QED: The Strange Theory of Light and Matter*. Princeton, NJ: Princeton University Press, 2006. First published in 1985.

Foster, Hal. "The ABCs of Contemporary Design." *October* 100, Spring 2002, 191–199.

Gannon, Todd, ed. *The Light Construction Reader*. New York: Monacelli Press, 2002.

Gropius, Walter. *The New Architecture and the Bauhaus*. Cambridge, Mass.: MIT Press, 1965.

Holl, Steven. *Questions of Perception: Phenomenology of Architecture*. San Francisco: William Stout, 2006. First published by A+U Publishing Co., 1994.

Korn, Arthur. *Glass in Modern Architecture*. London: Barrie & Rockcliff, 1968. First published in German as *Glas im Bau und als Gebrauchsgegenstand*, 1929.

Le Corbusier. *Towards a New Architecture*. New York: Dover, 1986. First published in French as *Vers un Architecture*, 1923.

Mertins, Detlef. *Modernity Unbound: Other Histories of Architectural Modernity*. London: AA Publications, 2011.

Plummer, Henry. *Masters of Light*. Tokyo: A+U Publishing Co., 2003.

Riley, Terence. *Light Construction*. New York: Museum of Modern Art, 1995.

Rowe, Colin and Robert Slutzky. "Transparency: Literal and Phenomenal," in *The Mathematics of the Ideal Villa and Other Essays*. Cambridge, Mass.: MIT Press, 1999. First published in *Perspecta* 8, 1963, 45–54.

Rowe, Colin and Robert Slutzky. "Transparency: Literal and Phenomenal, Part . . . 2." *Perspecta* 13/14, 1971, 287–301.

Twombly, Robert, ed. *Louis Kahn: Essential Texts*. New York: WW Norton, 2003.

Vidler, Anthony. *The Architectural Uncanny*. Cambridge, Mass.: MIT Press, 1992.

1 SOLIDIFIED LIGHT

Bernstein, Fred A. "Where Old and New Collide." *Metropolis*, February 2006, 114–117.

Frampton, Kenneth. "Maison de Verre." *Perspecta* 12, 1969, 77–126.

Holl, Steven. *Urbanisms: Working With Doubt*. New York: Princeton Architectural Press, 2009.

Lecuyer, Annette. "Artful Addition." *Architectural Review*, February 2006, 54–57.

Ouroussoff, Nicolai. "The Best House in Paris." *New York Times*, August 26, 2007, Arts & Leisure, 1, 22–23.

"Steven Holl Architects 2004–2008." *El Croquis* 141, 2008.

Vellay, Marc. *La Maison de Verre*. Tokyo: ADA Edita, 1988.

Vellay, Marc and Kenneth Frampton. *Pierre Chareau, Architect and Craftsman, 1883–1950*. New York: Rizzoli, 1986.

2 ART HOUSE CINEMA

Bee, Harriet, and Michelle Elligott, ed. *Art In Our Time: A Chronicle of the Museum of Modern Art*. New York: Museum of Modern Art, 2004.

Ingersoll, Richard. "Light Boxes." *Architecture*, October 1997, 90–101.

Lowry, Glenn D. *The New Museum of Modern Art*. New York: Museum of Modern Art, 2005.

"Museum of Modern Art, New York City." *Architectural Forum*, August 1939, 115–128.

"Mystical Presence." *Architectural Review*, December 1997, 46–53.

BIBLIOGRAPHY

Newhouse, Victoria. *Towards a New Museum.* New York: Monacelli Press, 2006.

Ricciotti, Dominic. "The 1939 Building of the Museum of Modern Art: The Goodwin–Stone Collaboration." *American Art Journal* 66, No. 3, Summer 1985, 50–76.

Stern, Robert A.M., Gregory Gilmartin, and Thomas Mellins. *New York 1930: Architecture and Urbanism Between the Two World Wars.* New York: Rizzoli, 1987.

Wojtowicz, Robert, ed. *Sidewalk Critic: Lewis Mumford's Writings on New York.* New York: Princeton Architectural Press, 1998.

Yoshida, Nobuyuki, ed. "Peter Zumthor." *A+U Architecture and Urbanism*, February 1998, Extra Edition.

Zumthor, Peter. *Kunsthaus Bregenz.* Ostfildern: Hatje, 1999.

3 CRYSTALLIZATION

Fernández-Galiano, Luis, ed. *Herzog & de Meuron 1978–2007.* Madrid: Arquitectura Viva, 2007.

Fissabre, Anke and Bernhard Niethammer. "The Invention of the Glazed Curtain Wall in 1903—The Steiff Toy Factory." *Proceedings of the Third International Congress on Construction History,* Cottbus, May 2009.

Gössel, Peter and Gabriele Leuthäuser. *Architecture in the Twentieth Century.* Cologne: Benedikt Taschen, 1991.

"Laban Centre in London." *Detail,* 7/8 2003, 791–797.

Meyer, Henning. "Historischer Fabrikbau der Firma Steiff in Giengen." *Baumeister,* November 2003, 90–95.

Reid, Giles. "Dancing Architecture." *Domus,* April 2003, 62–79.

4 COMPOUND LENS

Barreneche, Raul A. "Fashion House Hermès' Flagship Store by Renzo Piano," *Architectural Record,* October 2002, 166–170.

"Building for Corning Glass Works, New York City." *Architectural Forum,* December 1937, 457–460.

Gray, Christopher. "An Architectural Barometer." *New York Times,* September 1, 2011.

Herrick, George. "A Glass-Faced Building, New York." *Builder,* August 20, 1937, 314–315.

Jodidio, Philip. *Piano: Renzo Piano Building Workshop, 1966 to Today.* Cologne: Taschen, 2008.

"Kaufhaus in Tokio." *Detail,* July 2001, 1258–1261.

Mumford, Lewis. "The Sky Line: Gardens and Glass." *The New Yorker,* October 16, 1937, 68–70.

Neumann, Dietrich. "The Century's Triumph in Lighting: The Luxfer Prism Companies and Their Contribution to Early Modern Architecture." *Journal of the Society of Architectural Historians* 54, No. 1, March 1995, 24–53.

5 GEOLOGY

"Bibliothèque á l'Université de Yale, États-Unis." *Architecture d'aujourd'hui* 34, No. 117, Nov. 1964–Jan. 1965, 62–67.

Bognar, Botond. *Material Immaterial: The New Work of Kengo Kuma.* New York: Princeton Architectural Press, 2009.

Krinsky, Carol Herselle. *Gordon Bunshaft of Skidmore, Owings & Merrill.* Cambridge, Mass.: MIT Press, 1988.

Mateo, Josep Lluís. "New Headquarters for the Deutsche Bundesbank, Chemnitz, Germany 2004." *A+U Architecture and Urbanism* 447, December 2007, 24–31.

"Rare Book Library at Yale Dedicated." *Architectural Record* 133, November 1963, 12–13.

Verona, Irina. "Engineered Surfaces: Toward a Technology of Image." *Praxis* 9, 2007, 96–111.

6 BIOLUMINESCENCE

Beth Sholom Congregation. *Beth Sholom Synagogue: An American Synagogue* (pamphlet). Elkins Park, Penn., January 2008.

"Beth Sholom Synagogue, Philadelphia." *Architectural Record,* July 1954, 20.

Cohn, David. "Like Two Glowing Crystals, Rafael Moneo's Centro Kursaal." *Architectural Record,* May 2000, 212–223.

Davis, Patricia Talbot. *Together They Built a Mountain*. Lititz, Penn.: Sutter House, 1974.

Klein, Julia M. "The Rabbi and Frank Lloyd Wright." *Wall Street Journal*, December 22, 2009.

Moneo, Rafael. *The Freedom of the Architect*. Ann Arbor: University of Michigan, 2002.

Moneo, Rafael. *Remarks on 21 Works*. New York: Monacelli Press, 2010.

Patterson, Terry L. *Frank Lloyd Wright and the Meaning of Materials*. New York: Van Nostrand Reinhold, 1994.

"Rafael Moneo 1995–2000." *El Croquis* 98, 1999.

Hertzberg, Mark. *Frank Lloyd Wright's SC Johnson Research Tower*. San Francisco: Pomegranate, 2010.

Lipman, Jonathan. *Frank Lloyd Wright and the Johnson Wax Buildings*. Mineola, New York: Dover, 2003.

Martin, Reinhold. "The Crystal World: Frank Gehry's IAC." *Harvard Design Magazine* 27, Fall 2007/Winter 2008, 4–13.

Ouroussoff, Nicolai. "Gehry's New York Debut: Subdued Tower of Light." *New York Times*, March 22, 2007.

Pearson, Clifford. "InterActiveCorp Building, New York City." *Architectural Record*, October 2007, 112–119.

Tafel, Edgar. *Years with Frank Lloyd Wright: Apprentice to Genius*. New York: Dover, 1979.

7 FADE TO BLACK

Banham, Reyner. *Age of the Masters: A Personal View of Modern Architecture*. New York: Harper & Row, 1962.

Illustration Credits

All drawings are by the author and are intended to present an interpretation of the details of each project, based upon information that is publicly available or that has been provided by the architects. Drawings are intended to convey the general concept and arrangement of components in each design and should not be assumed to represent either the architects' construction documents or as-built conditions.

Figure 0.1 Transparency, Translucency, Opacity. Diagram by Scott Murray.

Figure 0.2 Bauhaus Building, Dessau, Germany. Walter Gropius, 1926. Photograph © Cethegus.

Figure 0.3 Entry hall. Tugendhat House, Brno, Czech Republic. Ludwig Mies van der Rohe, 1930. Photograph © Roberto Schezen / Esto.

Figure 0.4 Literal transparency. Design Research Store, Cambridge, Massachusetts, USA. Benjamin Thompson and Associates, 1969. Photograph by Ezra Stoller © Esto.

Figure 0.5 Shoren-in Temple, Kyoto, Japan. Late thirteenth century, rebuilt 1895. Photograph © Scott Murray.

Figure 0.6 Glass Pavilion, Toledo Museum of Art, Ohio, USA. SANAA, 2006. Photograph © Scott Murray.

Figure 0.7 Nelson-Atkins Museum of Art, Kansas City, Missouri, USA. Steven Holl Architects, 2007. Photograph © Scott Murray.

Figure 0.8 Nelson-Atkins Museum of Art, Kansas City, Missouri, USA. Steven Holl Architects, 2007. Photograph © Scott Murray.

Figure 1.1 Higgins Hall Insertion, Brooklyn, New York, USA. Steven Holl Architects, 2005. Photograph © David Sunberg / Esto.

Figure 1.2 Maison de Verre, Paris, France. Pierre Chareau and Bernard Bijvoet, 1932. Photograph © Architectural Press Archive / RIBA Library Photographs Collection.

Figure 1.3 Maison de Verre, Paris, France. Pierre Chareau and Bernard Bijvoet, 1932. Photograph © Scott Murray.

Figure 1.4 Maison de Verre, Paris, France. Pierre Chareau and Bernard Bijvoet, 1932. Photograph © Scott Murray.

Figure 1.5 Maison de Verre, Paris, France. Pierre Chareau and Bernard Bijvoet, 1932. Photograph © Emmanuel Thirard / RIBA Library Photographs Collection.

Figure 1.6 Typical glass lens. Maison de Verre, Paris, France. Pierre Chareau and Bernard Bijvoet, 1932. Photograph © Scott Murray.

Figure 1.7 Glass-lens curtain wall. Maison de Verre. Drawing by Scott Murray.

Figure 1.8 Typical module. Maison de Verre, Paris, France. Pierre Chareau and Bernard Bijvoet, 1932. Photograph © Scott Murray.

Figure 1.9 Interior. Maison de Verre, Paris, France. Pierre Chareau and Bernard Bijvoet, 1932. Photograph © Scott Murray.

Figure 1.10 Maison de Verre, Paris, France. Pierre Chareau and Bernard Bijvoet, 1932. Photograph © Scott Murray.

Figure 1.11 Maison de Verre, Paris, France. Pierre Chareau and Bernard Bijvoet, 1932. Photograph © Scott Murray.

Figure 1.12 Maison de Verre, Paris, France. Pierre Chareau and Bernard Bijvoet, 1932. Photograph © Scott Murray.

Figure 1.13 Higgins Hall Insertion, Brooklyn, New York, USA. Steven Holl Architects, 2005. Photograph © David Sunberg / Esto.

Figure 1.14 Higgins Hall Insertion, Brooklyn, New York, USA. Steven Holl Architects, 2005. Photograph © Scott Murray.

Figure 1.15 Higgins Hall Insertion, Brooklyn, New York, USA. Steven Holl Architects, 2005. Photograph © David Sunberg / Esto.

Figure 1.16 Studio interior. Higgins Hall Insertion, Brooklyn, New York, USA. Steven Holl Architects, 2005. Photograph © David Sunberg / Esto.

Figure 1.17 Dissonant zone. Higgins Hall Insertion, Brooklyn, New York, USA. Steven Holl Architects, 2005. Photograph © Scott Murray.

Figure 1.18 Interior. Higgins Hall Insertion, Brooklyn, New York, USA. Steven Holl Architects, 2005. Photograph © Scott Murray.

Figure 1.19 Channel-glass wall assembly. Higgins Hall Insertion. Drawing by Scott Murray.

Figure 1.20 Entrance lobby. Nelson-Atkins Museum of Art, Kansas City, Missouri, USA. Steven Holl Architects, 2007. Photograph © Scott Murray.

Figure 1.21 Higgins Hall Insertion, Brooklyn, New York, USA. Steven Holl Architects, 2005. Photograph © Scott Murray.

Figure 1.22 Higgins Hall Insertion, Brooklyn, New York, USA. Steven Holl Architects, 2005. Photograph © Scott Murray.

Figure 2.1 Kunsthaus Bregenz, Bregenz, Austria. Peter Zumthor, 1997. Photograph © Scott Murray.

Figure 2.2 Kunsthaus Bregenz on the shore of Lake Constance, Bregenz, Austria. Photograph by Markus Tretter © Siegrun Appelt KUB.

Figure 2.3 Kunsthaus Bregenz, Bregenz, Austria. Peter Zumthor, 1997. Photograph © Hélène Binet.

Figure 2.4 Kunsthaus Bregenz, Bregenz, Austria. Peter Zumthor, 1997. Photograph © Scott Murray.

Figure 2.5 Acid-etched laminated glass. Kunsthaus Bregenz, Bregenz, Austria. Peter Zumthor, 1997. Photograph © Scott Murray.

Figure 2.6 Outer-skin assembly. Kunsthaus Bregenz. Drawing by Scott Murray.

Figure 2.7 Stainless-steel angle supporting glass panel. Kunsthaus Bregenz, Bregenz, Austria. Peter Zumthor, 1997. Photograph © Scott Murray.

Figure 2.8 Gallery interior. Kunsthaus Bregenz, Bregenz, Austria. Peter Zumthor, 1997. Photograph © Scott Murray.

Figure 2.9 Kunsthaus Bregenz, Bregenz, Austria. Peter Zumthor, 1997. Photograph © Scott Murray.

Figure 2.10 Between the two skins. Kunsthaus Bregenz, Bregenz, Austria. Peter Zumthor, 1997. Photograph © Hélène Binet.

Figure 2.11 Kunsthaus Bregenz, Bregenz, Austria. Peter Zumthor, 1997. Photograph © Scott Murray.

Figure 2.12 Kunsthaus Bregenz, Bregenz, Austria. Peter Zumthor, 1997. Photograph © Scott Murray.

Figure 2.13 Kunsthaus Bregenz, Bregenz, Austria. Peter Zumthor, 1997. Photograph © Scott Murray.

Figure 2.14 Kunsthaus Bregenz, Bregenz, Austria. Peter Zumthor, 1997. Photograph © Hélène Binet.

Figure 2.15 Kirchner Museum, Davos, Switzerland. Gigon and Guyer, 1992. Photograph © Scott Murray.

Figure 2.16 Kirchner Museum, Davos, Switzerland. Gigon and Guyer, 1992. Photograph © Scott Murray.

Figure 2.17 Auguste Rodin Museum, Seoul, Korea. KPF Associates, 1999. Photograph © Ken McCown.

Figure 2.18 Mori Arts Center, Tokyo, Japan. Gluckman Mayner Architects, 2003. Photograph © Scott Murray.

Figure 2.19 Figge Art Museum, Davenport, Iowa, USA. David Chipperfield, 2005. Photograph © Scott Murray.

Figure 2.20 Nelson-Atkins Museum of Art, Kansas City, Missouri, USA. Steven Holl Architects, 2007. Photograph © Scott Murray.

Figure 2.21 Museum of Modern Art, New York City, USA. Goodwin and Stone, 1939. Photograph © Museum of the City of New York.

Figure 2.22 Construction photo, 12 September 1938, Museum of Modern Art, New York City, USA. Goodwin and Stone. Photograph © The Museum of Modern Art / Licensed by SCALA / Art Resource, NY.

Figure 2.23 Museum of Modern Art, New York City, USA. Goodwin and Stone, 1939. Photograph by Wurts Bros. © Museum of the City of New York.

Figure 2.24 Interior staircase and Thermolux curtain wall. Museum of Modern Art, New York City, USA. Goodwin and Stone, 1939. Photograph © The Museum of Modern Art / Licensed by SCALA / Art Resource, NY.

Figure 2.25 Thermolux curtain wall. Museum of Modern Art. Drawing by Scott Murray.

Figure 2.26 Restored Thermolux curtain wall, interior, 2004. Museum of Modern Art. Photograph © Scott Murray.

Figure 2.27 Restored Thermolux curtain wall, interior, 2004. Museum of Modern Art. Photograph © Scott Murray.

Figure 2.28 Restored façade with Thermolux curtain wall, 2004. Museum of Modern Art. Photograph © Scott Murray.

Figure 3.1 Steiff Factory Building, Giengen an der Brenz, Germany. Richard Steiff, 1903. Photograph © Margarete Steiff GmbH.

Figure 3.2 Aerial view of the Steiff Factory complex, 1910. Photograph © Margarete Steiff GmbH.

Figure 3.3 Extension of the Steiff Factory, 1904. Photograph © Margarete Steiff GmbH.

Figure 3.4 Construction of the Steiff Factory Building, 1903. Photograph © Margarete Steiff GmbH.

Figure 3.5 Double-skin curtain wall. Steiff Factory Building. Drawing by Scott Murray.

Figure 3.6 Interior, Steiff Factory Building, Giengen an der Brenz, Germany. Richard Steiff, 1903. Photograph © Margarete Steiff GmbH.

Figure 3.7 Steiff Factory Building, Giengen an der Brenz, Germany. Richard Steiff, 1903. Photograph © Margarete Steiff GmbH.

Figure 3.8 CIBA Plant and Laboratories, Duxford, England. Ove Arup and Partners, 1958. Photograph © Colin Westwood / RIBA Library Photographs Collection.

Figure 3.9 Laban Dance Center, Deptford, London, UK. Herzog & de Meuron, 2003. Photograph © Scott Murray.

Figure 3.10 Laban Dance Center, Deptford, London, UK. Herzog & de Meuron, 2003. Photograph © Scott Murray.

Figure 3.11 Laban Dance Center, Deptford, London, UK. Herzog & de Meuron, 2003. Photograph © Scott Murray.

Figure 3.12 Tate Modern Gallery, London, UK. Herzog & de Meuron, 2000. Photograph © Scott Murray.

Figure 3.13 De Young Museum, San Francisco, California, USA. Herzog & de Meuron, 2005. Photograph © Scott Murray.

Figure 3.14 Tinted polycarbonate panels design in collaboration with the artist Michael Craig-Martin. Laban Dance Center, Deptford, London, UK. Herzog & de Meuron, 2003. Photograph © Scott Murray.

Figure 3.15 Limoges Concert Hall, Limoges, France. Bernard Tschumi Architects, 2007. Photograph © Peter Mauss / Esto.

Figure 3.16 Dance studio interior. Laban Dance Center, Deptford, London, UK. Herzog & de Meuron, 2003. Photograph © Merlin Hendy.

Figure 3.17 Laban Dance Center, Deptford, London, UK. Herzog & de Meuron, 2003. Photograph © Daniel Hewitt / RIBA Library Photographs Collection.

Figure 3.18 Double-skin wall assembly. Laban Dance Center. Drawing by Scott Murray.

Figure 3.19 Dance studio interior. Laban Dance Center, Deptford, London, UK. Herzog & de Meuron, 2003. Photograph © View Pictures Ltd / Super Stock.

Figure 3.20 Laban Dance Center, Deptford, London, UK. Herzog & de Meuron, 2003. Photograph © Scott Murray.

Figure 3.21 Laban Dance Center, Deptford, London, UK. Herzog & de Meuron, 2003. Photograph © Scott Murray.

Figure 4.1 Maison Hermès, Tokyo, Japan. Renzo Piano Building Workshop, 2001. Photograph © Scott Murray.

Figure 4.2 New York Times Tower, New York City, USA. Renzo Piano Building Workshop, 2007. Photograph © Scott Murray.

Figure 4.3 Harumi Dori, Ginza district, Tokyo, Japan. Maison Hermès at center. Photograph © Scott Murray.

Figure 4.4 Maison Hermès, Tokyo, Japan. Renzo Piano Building Workshop, 2001. Photograph © Scott Murray.

Figure 4.5 Glass-block curtain wall. Maison Hermès. Drawing by Scott Murray.

Figure 4.6 Maison Hermès, Tokyo, Japan. Renzo Piano Building Workshop, 2001. Photograph © Scott Murray.

Figure 4.7 Maison Hermès, Tokyo, Japan. Renzo Piano Building Workshop, 2001. Photograph © Scott Murray.

Figure 4.8 Interior, Maison Hermès, Tokyo, Japan. Renzo Piano Building Workshop, 2001. Photograph © Age Fotostock / Super Stock.

Figure 4.9 Maison Hermès, Tokyo, Japan. Renzo Piano Building Workshop, 2001. Photograph © Scott Murray.

Figure 4.10 Maison Hermès, Tokyo, Japan. Renzo Piano Building Workshop, 2001. Photograph © Scott Murray.

Figure 4.11 Maison Hermès, Tokyo, Japan. Renzo Piano Building Workshop, 2001. Photograph © Scott Murray.

Figure 4.12 Academy of Arts, Maastricht, the Netherlands. Wiel Arets Architects, 1993. Photograph © Jan Bitter.

Figure 4.13 Barclay Simpson Sculpture Studio, Oakland, California, USA. Jim Jennings, 1992. Photograph © Scott Murray.

Figure 4.14 Christian Dior Building, Tokyo, Japan. SANAA, 2003. Photograph © Scott Murray.

Figure 4.15 Louis Vuitton Store, Tokyo, Japan. Jun Aoki, 2004. Photograph © Scott Murray.

Figure 4.16 Fifth Avenue façade, Steuben Glass Building, New York City, USA. William and Geoffrey Platt, 1937. Photograph by Gottscho-Schleisner © Museum of the City of New York.

Figure 4.17 Steuben Glass Building, New York City, USA. William and Geoffrey Platt, 1937. Library of Congress, Prints and Photographs Division, Gottscho-Schleisner Collection.

Figure 4.18 Glass-block wall assembly. Steuben Glass Building. Drawing by Scott Murray.

Figure 4.19 Office interior, Steuben Glass Building, New York City, USA. William and Geoffrey Platt, 1937. Photograph by Gottscho-Schleisner © Museum of the City of New York.

Figure 4.20 56th Street façade, Steuben Glass Building, New York City, USA. William and Geoffrey Platt, 1937. Photograph by Gottscho-Schleisner © Museum of the City of New York.

Figure 4.21 Corning Glass Tower, New York City, USA. Wallace K. Harrison, 1959. Note Steuben Glass Building at right. Photograph by Ezra Stoller © Esto.

Figure 4.22 Corning Glass Tower. Note renovated façade of Harry Winston Building (formerly Steuben Glass Building) at right. Photograph by Ezra Stoller © Esto.

Figure 4.23 Harry Winston Building (formerly Steuben Glass Building). Photograph © Scott Murray.

Figure 4.24 Lescaze Townhouse and Office, New York City, USA. William Lescaze, 1934. Photograph © Museum of the City of New York.

Figure 4.25 Glass Pavilion, Werkbund Exhibition, Cologne, Germany. Bruno Taut, 1914. Photograph © Foto Marburg / Art Resource, NY.

Figure 5.1 Beinecke Rare Book Library, New Haven, Connecticut, USA. Gordon Bunshaft, SOM, 1963. Photograph © Scott Murray.

Figure 5.2 Manufacturers Hanover Trust Company, New York City, USA. Gordon Bunshaft, SOM, 1954. Photograph by Ezra Stoller © Esto.

Figure 5.3 Beinecke Rare Book Library, New Haven, Connecticut, USA. Gordon Bunshaft, SOM, 1963. Photograph © Scott Murray.

Figure 5.4 Interior view from mezzanine level. Beinecke Rare Book Library, New Haven, Connecticut, USA. Gordon Bunshaft, SOM, 1963. Photograph © Wolfgang Hoyt / Esto.

Figure 5.5 Overlooking the courtyard from plaza level. Beinecke Rare Book Library, New Haven, Connecticut, USA. Gordon Bunshaft, SOM, 1963. Photograph © Scott Murray.

Figure 5.6 Wall assembly. Beinecke Rare Book Library. Drawing by Scott Murray.

Figure 5.7 Construction of the Beinecke Rare Book Library.

Photograph courtesy of the Beinecke Rare Book and Manuscript Library, Yale University.

Figure 5.8 Interior view of marble cladding. Beinecke Rare Book Library. Photograph © Scott Murray.

Figure 5.9 Interior view of marble cladding at mezzanine. Beinecke Rare Book Library. Photograph © Scott Murray.

Figure 5.10 Preliminary model of the Beinecke Rare Book Library. Photograph by Ezra Stoller © Esto

Figure 5.11 Beinecke Rare Book Library, New Haven, Connecticut, USA. Gordon Bunshaft, SOM, 1963. Photograph © Scott Murray.

Figure 5.12 Beinecke Rare Book Library, New Haven, Connecticut, USA. Gordon Bunshaft, SOM, 1963. Photograph © Scott Murray.

Figure 5.13 Amoco Building (now Aon Center), Chicago, Illinois, USA. Edward Durell Stone, 1973. Photograph © Scott Murray.

Figure 5.14 St. Pius Church, Meggen, Switzerland. Franz Fueg, 1966. Photograph © Leiju.

Figure 5.15 Cathedral of Our Lady of the Angels, Los Angeles, California, USA. Rafael Moneo, 2002. Photograph © Scott Murray.

Figure 5.16 LVMH Building, Osaka, Japan. Kengo Kuma and Associates, 2004. Photograph © Scott Murray.

Figure 5.17 LVMH Building, Osaka, Japan. Kengo Kuma and Associates, 2004. Photograph © Scott Murray.

Figure 5.18 LVMH Building, Osaka, Japan. Kengo Kuma and Associates, 2004. Photograph © Kengo Kuma and Associates.

Figure 5.19 LVMH Building, Osaka, Japan. Kengo Kuma and Associates, 2004. Photograph © Scott Murray.

Figure 5.20 Exterior view of illuminated curtain wall. LVMH Building, Osaka, Japan. Kengo Kuma and Associates, 2004. Photograph © Scott Murray.

Figure 5.21 Interior view of curtain wall. LVMH Building, Osaka, Japan. Kengo Kuma and Associates, 2004. Photograph © Kengo Kuma and Associates.

Figure 5.22 LVMH Building, Osaka, Japan. Kengo Kuma and Associates, 2004. Photograph © Scott Murray.

Figure 5.23 Curtain-wall mullion detail. LVMH Building, Osaka. Drawing by Scott Murray.

Figure 5.24 LVMH Building, Osaka, Japan. Kengo Kuma and Associates, 2004. Photograph © Scott Murray.

Figure 5.25 Asahi Broadcasting Corporation Headquarters, Osaka, Japan. Kengo Kuma and Associates, 2008. Photograph © Scott Murray.

Figure 5.26 Tiffany Flagship Store, Ginza, Tokyo, Japan. Kengo Kuma and Associates, 2008. Photograph © Scott Murray.

Figure 5.27 Deutsche Bundesbank Headquarters, Chemnitz, Germany. Josep Lluis Mateo, 2004. Photograph © Jan Bitter.

Figure 6.1 Kursaal Congress Center, San Sebastián, Spain. Rafael Moneo, 1999. Photograph © Scott Murray.

Figure 6.2 View from Playa Zurriola, day. Kursaal Congress Center, San Sebastián, Spain. Rafael Moneo, 1999. Photograph © Scott Murray.

Figure 6.3 View from Playa Zurriola, night. Kursaal Congress Center, San Sebastián, Spain. Rafael Moneo, 1999. Photograph © Scott Murray.

Figure 6.4 Concave translucent laminated glass. Kursaal Congress Center, San Sebastián, Spain. Rafael Moneo, 1999. Photograph © Scott Murray.

Figure 6.5 Double-skin curtain wall. Kursaal Congress Center. Drawing by Scott Murray.

Figure 6.6 View from Playa Zurriola, night. Kursaal Congress Center, San Sebastián, Spain. Rafael Moneo, 1999. Photograph © Scott Murray.

Figure 6.7 Kursaal Congress Center, San Sebastián, Spain. Rafael Moneo, 1999. Photograph © Scott Murray.

Figure 6.8 Interior. Kursaal Congress Center, San Sebastián, Spain. Rafael Moneo, 1999. Photograph © Scott Murray.

Figure 6.9 Interior, framed view toward Playa Zurriola. Kursaal Congress Center, San Sebastián, Spain. Rafael Moneo, 1999. Photograph © Scott Murray.

Figure 6.10 Interior. Kursaal Congress Center, San Sebastián, Spain. Rafael Moneo, 1999. Photograph © Scott Murray.

Figure 6.11 55 Water Street Plaza, New York City, USA. Rogers Marvel Architects and Ken Smith Landscape Architect, 2005. Photograph © Francis Dzikowski / Esto.

Figure 6.12 Dewey Square T-Station, Boston, Massachusetts, USA. Machado and Silvetti Associates, 2007. Photograph © Anton Grassl / Esto.

Figure 6.13 Kursaal Congress Center, San Sebastián, Spain. Rafael Moneo, 1999. Photograph © Scott Murray.

Figure 6.14 Beth Sholom Synagogue, Elkins Park, Pennsylvania, USA. Frank Lloyd Wright, 1959. Library of Congress, Prints and Photographs Division, HABS PA, 46-ELKPA, 1-1.

Figure 6.15 Beth Sholom Synagogue, Elkins Park, Pennsylvania, USA. Frank Lloyd Wright, 1959. Photograph © Scott Murray.

Figure 6.16 Interior. Beth Sholom Synagogue, Elkins Park, Pennsylvania, USA. Frank Lloyd Wright, 1959. Photograph © Scott Murray.

Figure 6.17 Beth Sholom Synagogue, Elkins Park, Pennsylvania, USA. Frank Lloyd Wright, 1959. Photograph © Scott Murray.

Figure 6.18 Sanctuary interior. Beth Sholom Synagogue, Elkins Park, Pennsylvania, USA. Frank Lloyd Wright, 1959. Photograph © Scott Murray.

Figure 6.19 Beth Sholom Synagogue, Elkins Park, Pennsylvania, USA. Frank Lloyd Wright, 1959. Photograph © Scott Murray.

Figure 6.20 Double-skin envelope assembly. Beth Sholom Synagogue. Drawing by Scott Murray.

Figure 6.21 Interior view of double-skin envelope. Beth Sholom Synagogue, Elkins Park, Pennsylvania, USA. Frank Lloyd Wright, 1959. Photograph © Scott Murray.

Figure 6.22 Interior view of wall with integrated light fixture. Beth Sholom Synagogue, Elkins Park, Pennsylvania, USA. Frank Lloyd Wright, 1959. Photograph © Scott Murray.

Figure 6.23 Interior view of ceiling. Beth Sholom Synagogue, Elkins Park, Pennsylvania, USA. Frank Lloyd Wright, 1959. Photograph © Scott Murray.

Figure 6.24 History Faculty Building, Cambridge University, Cambridge, UK. James Stirling, 1967. Photograph by Ezra Stoller © Esto.

Figure 6.25 Prayer Pavilion of Light, Phoenix, Arizona, USA. DeBartolo Architects, 2007. Photograph © Bill Timmerman.

Figure 6.26 Interior. Beth Sholom Synagogue, Elkins Park, Pennsylvania, USA. Frank Lloyd Wright, 1959. Photograph © Scott Murray.

Figure 7.1 InterActiveCorp Headquarters, New York City, USA. Gehry Partners, 2007. Photograph © Scott Murray.

Figure 7.2 InterActiveCorp Headquarters, New York City, USA. Gehry Partners, 2007. Photograph © Scott Murray.

Figure 7.3 Vitra Design Museum, Weil am Rhein, Germany. Gehry Partners, 1989. Photograph © Scott Murray.

Figure 7.4 Vontz Center for Molecular Studies, Cincinnati, Ohio, USA. Gehry Partners, 1999. Photograph © Scott Murray.

Figure 7.5 Center for the Visual Arts, Toledo Museum of Art, Ohio, USA. Gehry Partners, 1992. Photograph © Scott Murray.

Figure 7.6 Walt Disney Concert Hall, Los Angeles, California, USA. Gehry Partners, 2003. Photograph © Scott Murray.

Figure 7.7 InterActiveCorp Headquarters, New York City, USA. Gehry Partners, 2007. Photograph © Scott Murray.

Figure 7.8 Close-up exterior view of ceramic frit dot pattern on curtain-wall glass. InterActiveCorp Headquarters, New York City, USA. Gehry Partners, 2007. Photograph © Scott Murray.

Figure 7.9 B3 Office Building, Stockley Park, London, UK. Norman Foster, 1989. Photograph © Joe Low / RIBA Library Photographs Collection.

Figure 7.10 Louis Vuitton Building, New York City, USA. Jun Aoki, 2004. Photograph © Scott Murray.

Figure 7.11 Spertus Institute of Jewish Studies, Chicago, Illinois, USA. Krueck + Sexton Architects, 2007. Photograph © Scott Murray.

Figure 7.12 Utrecht University Library, the Netherlands. Wiel Arets Architects, 2004. Photograph © Scott Murray.

Figure 7.13 IKMZ Building, Cottbus University, Germany. Herzog & de Meuron, 2004. Photograph © Alexandru Giurca.

Figure 7.14 InterActiveCorp Headquarters, New York City, USA. Gehry Partners, 2007. Photograph © Scott Murray.

Figure 7.15 Unitized curtain-wall system. InterActiveCorp Headquarters. Drawing by Scott Murray.

Figure 7.16 Installation of curtain wall. InterActiveCorp Headquarters, New York City, USA. Gehry Partners, 2007. Photograph © Scott Murray.

Figure 7.17 InterActiveCorp Headquarters, New York City, USA. Gehry Partners, 2007. Photograph © Albert Vecerka / Esto.

Figure 7.18 Office interior. InterActiveCorp Headquarters, New York City, USA. Gehry Partners, 2007. Photograph © Albert Vecerka / Esto.

Figure 7.19 Lobby interior, day. InterActiveCorp Headquarters, New York City, USA. Gehry Partners, 2007. Photograph © Scott Murray.

Figure 7.20 Lobby exterior, night. InterActiveCorp Headquarters, New York City, USA. Gehry Partners, 2007. Photograph © Albert Vecerka / Esto.

Figure 7.21 Johnson Wax Research Tower, Racine, Wisconsin, USA. Frank Lloyd Wright, 1950. Photograph by Ezra Stoller © Esto.

Figure 7.22 Johnson Wax Research Tower, left, 1950, and Administration Building, right, 1939, Racine, Wisconsin, USA. Frank Lloyd Wright. Library of Congress, Prints and Photographs Division, HABS WIS, 51-RACI, 5-1.

Figure 7.23 Johnson Wax Administration Building, Racine, Wisconsin, USA. Frank Lloyd Wright, 1939. Library of Congress, Prints and Photographs Division, HABS WIS, 51-RACI, 5-1.

Figure 7.24 Johnson Wax Administration Building, Racine, Wisconsin, USA. Frank Lloyd Wright, 1939. Library of Congress, Prints and Photographs Division, HABS WIS, 51-RACI, 5-1.

Figure 7.25 Laboratory interior. Johnson Wax Research Tower, Racine, Wisconsin, USA. Frank Lloyd Wright, 1950. Photograph by Ezra Stoller © Esto.

Figure 7.26 Johnson Wax Research Tower, left, 1950, and Administration Building, right, 1939, Racine, Wisconsin, USA. Frank Lloyd Wright. Photograph © Scott Murray.

Figure 7.27 Glass wall assembly. Johnson Wax Research Tower. Drawing by Scott Murray.

Figure 7.28 Night view. Johnson Wax Research Tower, Racine, Wisconsin, USA. Frank Lloyd Wright, 1950. Photograph by Ezra Stoller © Esto.

Figure 7.29 Detail view, glass-tube wall. Johnson Wax Administration Building, Racine, Wisconsin, USA. Frank Lloyd Wright, 1939. Library of Congress, Prints and Photographs Division, HABS WIS, 51-RACI, 5-1.

Figure 7.30 Johnson Wax Research Tower, Racine, Wisconsin, USA. Frank Lloyd Wright, 1950. Library of Congress, Prints and Photographs Division, HABS WIS, 51-RACI, 5-1.

Figure 7.31 Johnson Wax Research Tower, left, 1950, and Administration Building, right, 1939, Racine, Wisconsin, USA. Frank Lloyd Wright. Photograph by Ezra Stoller © Esto.

Index

ND - #0039 - 090821 - C200 - 276/219/12 - PB - 9780415689311 - Gloss Lamination